Book Endorsement

Drawing on the diverse fields of landscape architecture, environmental history, art history, biology, and historical geography, this innovative and quirky collection analyses horticultural practice and design, heritage, material culture, and cultural representations of gardens from around the Asia-Pacific to break new ground in our understandings of the histories of gardens and designed landscapes. Here, seventeenth-century Chinese poetry joins a public park in fin de siècle New Zealand, while deep time accompanies the ubiquitous Chinese willow pattern. In what will be essential reading for scholars, practitioners and students, the authors offer fresh insights into the exciting cross-cultural possibilities for understanding gardens in their environmental and cultural contexts.

Dr. Ruth Morgan, Monash University

Gardens at the Frontier

Gardens at the Frontier addresses broad issues of interest to architectural historians, environmental historians, garden writers, geographers, and other scholars. It uses different disciplinary perspectives to explore garden history's thematic, geographical, and methodological frontiers through a focus on gardens as sites of cultural contact. The contributors address the extent to which gardens inhibit or further cultural contact; the cultural translation of garden concepts, practices and plants from one place to another; the role of non-written sources in cultural transfer; and which disciplines study gardens and designed landscapes, and how and why their approaches vary.

Chapters cover a range of designed landscapes and locations, periods and approaches: medieval Japanese *roji* (tea gardens); a seventeenth-century garden of southern China; post-war Australian 'natural gardens'; iconic twentieth-century American modernist gardens; 'international' willow-pattern design; geology and designed landscapes; gnomes; and landscape authorship of a public garden. Each chapter examines transfers of cultural ideas and their physical denouement.

This book was originally published as a special issue of *Studies in the History of Gardens & Designed Landscapes*.

James Beattie is Associate Professor of Science in Society at Victoria University of Wellington, New Zealand. His work focuses on the Asia-Pacific region, mostly over the last 200 years. He is especially interested in cross-cultural exchanges occasioned by British imperialism, and the nexus between environments, gardens, health, and art. He is author of nine books and over 60 articles and chapters.

Gardens at the Frontier

New Methodological Perspectives on Garden History and Designed Landscapes

Edited by
James Beattie

Routledge
Taylor & Francis Group

LONDON AND NEW YORK

First published 2018
by Routledge

2 Park Square, Milton Park, Abingdon, Oxfordshire OX14 4RN
52 Vanderbilt Avenue, New York, NY 10017

Routledge is an imprint of the Taylor & Francis Group, an informa business

First issued in paperback 2020

British Library Cataloguing in Publication Data
A catalogue record for this book is available from the British Library

ISBN 13: 978-0-8153-4790-3 (hbk)
ISBN 13: 978-0-367-58970-7 (pbk)

Typeset in Bembo
by RefineCatch Limited, Bungay, Suffolk

Publisher's Note
The publisher accepts responsibility for any inconsistencies that may have
arisen during the conversion of this book from journal articles to book chapters,
namely the possible inclusion of journal terminology.

Disclaimer
Every effort has been made to contact copyright holders for their permission to
reprint material in this book. The publishers would be grateful to hear from any
copyright holder who is not here acknowledged and will undertake to rectify
any errors or omissions in future editions of this book.

Contents

CONTENTS

Citation Information

The chapters in this book were originally published in *Studies in the History of Gardens & Designed Landscapes*. When citing this material, please use the original page numbering for each article, as follows:

Introduction
Gardens at the frontier: new methodological perspectives on garden history and designed landscapes: Introduction to Special Issue
James Beattie
Studies in the History of Gardens & Designed Landscapes, volume 36, issue 1 (January–March 2016), pp. 1–4

Chapter 1
Chinese Sources in the Japanese tea garden
Richard Bullen
Studies in the History of Gardens & Designed Landscapes, volume 36, issue 1 (January–March 2016), pp. 5–16

Chapter 2
China on a plate: a willow pattern garden realized
James Beattie
Studies in the History of Gardens & Designed Landscapes, volume 36, issue 1 (January–March 2016), pp. 17–31

Chapter 3
Zheng Yuanxun's 'A Personal Record of My Garden of Reflections'
Translated and Introduced by Duncan M. Campbell
Studies in the History of Gardens & Designed Landscapes, volume 29, issue 4 (April 2009), pp. 270–281

For any permission-related enquiries please visit:
http://www.tandfonline.com/page/help/permissions

Notes on Contributors

James Beattie is Associate Professor of Science in Society at Victoria University of Wellington, New Zealand. His work focuses on the Asia-Pacific region, mostly over the last 200 years. He is especially interested in cross-cultural exchanges occasioned by British imperialism, and the nexus between environments, gardens, health, and art. He is author of nine books and over 60 articles and chapters.

Jacky Bowring is Professor of Landscape Architecture at Lincoln University, New Zealand. Her research interests circle cultural landscape, history, memory, and emotion. She is the editor of the journal *Landscape Review*, and author of *Melancholy and the Landscape: Locating Sadness, Memory and Reflection in the Landscape* (2016).

Richard Bullen is Associate Professor of History of Art at the University of Canterbury, Christchurch, New Zealand. His principal areas of research are Japanese art and aesthetics, and Chinese art in New Zealand. With James Beattie, he is the editor of *New China Eyewitness: Roger Duff, Rewi Alley and the Art of Museum Diplomacy* (2017).

Duncan N. Campbell is a Teaching and Research Fellow at Victoria University of Wellington, New Zealand. His research and teaching focuses on the literary and material culture of late imperial China. More broadly, his research explores representations of private gardens in literature and art, private letter writing and diaries, late Imperial Chinese print culture, and auto/biographical writing.

Ian C. Duggan is Senior Lecturer in Biological Sciences at the University of Waikato, New Zealand. His main interests are invasion biology and zooplankton ecology. In particular, he is interested in the exploration of biological invasion vectors and pathways responsible for transportation of species at global or finer scales.

Christina Dyson is a Senior Heritage Consultant for Context Heritage Consultants, based in Melbourne, Australia. She specializes in twentieth-century Australian plant gardens, and has wide experience in Australia's urban and built heritage. She holds a PhD from the Faculty of Architecture, Building and Planning at the University of Melbourne.

Ian Henderson taught garden design, garden history, and indigenous landscapes in the Department of Landscape Architecture at UNITEC, Auckland, New Zealand, for 19 years. He is the founder (with Penny Cliffin) of HIKOI Garden Tours, which specializes in a professional approach to touring gardens with a strong emphasis on their cultural and historical context.

Michael M. Roche is Professor in the School of People, Environment and Planning at Massey University, New Zealand. His research interests include historical geography, especially colonial forestry and landscape transformation; contemporary agrifood studies; and the history of geographic thought, especially as it relates to New Zealand.

Preface

James Beattie

It is my pleasure to be able to present as an edited book an expanded version of a special issue of the same name that first appeared in *Studies in the History of Gardens & Designed Landscapes*, volume 36, issue 1 in March 2016. In addition to the original eight articles which appeared in the journal issue, the issue includes Duncan M. Campbell's "Zheng Yuanxun's 'A Personal Record of My Garden of Reflections,'" also from *Studies in the History of Gardens & Designed Landscapes*.

Campbell's moving translation and erudite analysis of Zheng Yuanxun's account concerns a garden Zheng had built in the Ming (1338–1644) city of Yangzhou, southern China, during the dying years of that troubled but culturally effervescent dynasty. Campbell's contribution strikes straight at the heart of several of the main themes of this edited book. The account illustrates the challenge of writing cultural histories of gardens whose physical traces have long since disappeared. Zheng provides details of a garden inspired by landscape painting—itself, of course, inseparable from calligraphy—and later immortalised both in Zheng's own account and in the celebrated poetry collection compiled on the occasion of 'the flowering of a single unusual yellow tree peony' in that garden. In this respect, Zheng's garden contrasts Michael Roche's chapter in the volume, which reassembles from the remaining physical traces a cultural history of a garden whose written record is mostly lost.

Zheng's record also speaks to the theme of cultural contact. It illustrates the manner in which other surviving gardens of the wider region (Jiangnan) to which Zheng's one belonged, provided contact zones enabling the new Manchu leaders of the conquering Qing dynasty (1644–1912) to learn important lessons about Han Chinese culture and values.[1] Even though it did not last much into the Qing, Zheng's 'Garden of Reflections' adds greatly to understandings of the cultural history of other Ming gardens which did survive into the new dynasty. Not least, as Campbell notes, Zheng's garden exemplifies commonly found tensions between the ideal of such gardens as sites of physical reclusion and the active role they played in social and political activities, a tension apparent in the various garden accounts translated by Campbell. The garden's design also manifests tensions of a different kind: 'between the physical constrictions of the site . . . and . . . the desire to achieve "understated elegance of the simple and rustic" . . . whilst avoiding any anxiety about the exhaustibility of the delights it contained as one moved through it.' This applies to many designed landscapes, but particularly so to gardens from this place and period of Chinese history: namely, the tension of having to create an allusion of naturalness from a site created through artifice. Finally, the inspiration for the 'Garden of Reflections' in Ji Cheng's (1582–c.1642) 'Craft of Gardens' (*Yuan Ye*, written circa 1631–1635) speaks to the singular importance of this text on garden design in China, and especially in more recent years overseas, through the proliferation of Jiangnan-style gardens as sites mediating cultural contact between Chinese and non-Chinese cultures, as well as among diverse Chinese cultural traditions.[2]

Since publication of the special journal issue in 2016, there have been some minor amendments to some of the papers. For Christina Dyson's chapter ("Rethinking Australian natural gardens and national identity, 1950–1979"), she would like to acknowledge Jean Walker (1922–2017), and also note that figures 3 and 4 were reproduced by permission of Elizabeth T. Smith and Ian and Rohan Walker. In his chapter, "W. W. Smith and the transformation of the Ashburton Domain 'from a wilderness into a beauty spot', 1894 to 1904," Michael Roche provides the following updated biography of W. W. Smith's early years. Smith was at Rose Hill, Carlisle, as under-gardener for Henry Lonsdale in 1871 and gardener in 1872–1874 at Burley-on-the-hill near Oakham, Rutland (the residence of George Henry Finch, the local MP), rather than the more famous Burghley Estate as stated in the chapter. Fortunately, this biographical amendment does not alter the chapter's argument. Roche is grateful to Dr Ross Galbreath for untangling Smith's early career movements and for the editor for enabling him to correct his original statement. Ian Henderson acknowledges permission to republish figure 1 in his chapter, by courtesy of the Estate of Ian Hamilton Finlay, and figures 3 and 5 by permission from the Thomas D. Church Collection, Environmental Design Archives, University of California, Berkeley, whose collection is the property of the Regents of the University of California. Since publication of the 2016 issue, James Beattie's affiliation has changed, to Victoria University of Wellington, New Zealand.

Additional acknowledgements

I would like to thank Duncan M. Campbell for allowing me to include his article in this special issue. It was originally published as: "Zheng Yuanxun's 'A Personal Record of My Garden of Reflections'", *Studies in the History of Gardens & Designed Landscapes*, 29:4, 2010, 270–281, DOI: 10.1080/14601170802426044. I also acknowledge the support of the Rachel Carson Center, Ludwigs-Maximilians University, Munich, where I was a Fellow while finalising this issue in August-September 2017.

Notes

1. This is taken up by Duncan M. Campbell in a forthcoming article in a special issue on 'Parks and Gardens in Environmental History', *Environment and History*, xxiv, 2018, edited by Karen Jones and James Beattie: Duncan M. Campbell, 'The Private Chinese Library as Contact Zone: The Little Mountain Hall Collection as Case Study', *Environment and History*, xxiv, 2018, pp. 103–19.
2. In China, *Yuan Ye* lay forgotten for centuries until its rediscovery and republication in China in 1931. Since then, its popularity has soared among designers, with its influence especially evident among the many dozens of Chinese gardens built overseas. For its ambiguous history and somewhat problematic representation in more recent years as emblematic of being 'the Chinese garden', see Craig Clunas, *Fruitful Sites: Garden Culture in Ming Dynasty China*, Durham; London, 1996. As an example of its more-recent overseas influence, see James Beattie and Duncan M. Campbell, *The Art, Culture and History of Lan Yuan* 蘭園: *The Dunedin Chinese Garden*, Dunedin; Shanghai, 2013.

Introduction
Gardens at the frontier: new methodological perspectives on garden history and designed landscapes

'Gardens at the Frontier: New Methodological Perspectives on Garden History and Designed Landscapes' brings together scholarly perspectives drawn from a variety of disciplines to explore garden history's thematic, geographical, and methodological frontiers through a focus on gardens as sites of cultural contact. Contributors address one or both of the following questions:

1. To what extent do gardens inhibit or further cultural contact?
2. What new methodological frontiers can examination of gardens at the frontier open up for scholarly perspectives on gardens and designed landscapes?

In thinking about these questions, contributors to this special issue consider the applicability to garden history of the concepts of cultural contact zone and cultural translation. In 1991, social and cultural analyst Mary Louise Pratt coined the term 'contact zone' to denote those 'social spaces where cultures meet, clash, and grapple with each other, often in contexts of highly asymmetrical relations of power, such as colonialism, slavery, or their aftermaths as they are lived out in many parts of the world today'.[1] Contributors also examine the broader significance to garden history of literary studies scholar David Porter's model of 'cultural translation'. Originally pertaining to the history of Chinese gardens in eighteenth-century Britain, cultural translation refers to the processes and adaptive strategies 'by which one culture finds meaning in another' through transfers of garden ideas, practices, and plants.[2]

Producing gardens and garden histories

Contributors also consider how past people and present garden researchers have produced gardens and garden histories. If geographers and architects are particularly strong at analysing space, historians at reading textual documents, and botanists at examining past garden-species distribution, what can approaches bringing together these — and other — disciplines offer for garden history?

The authors reflect on the production of gardens and their histories, in relation to the following questions:

- Which disciplines study gardens and designed landscapes, and how and why do their approaches vary?

- What sources do historians of gardens use and what methodologies can they employ to analyse such spaces?
- Can we translate particular concepts about gardens into another culture without losing their original immediacy or meaning?
- How does scale and temporality affect the impact and nature of garden history?
- Which other non-garden art forms and theories have informed the study of gardens, and how have garden studies benefited?
- What are the various textual and visual means of recording and writing garden history?

Papers

The seven papers of the special issue cover a range of designed landscapes, periods, and approaches: post-war Australian 'natural gardens'; medieval Japanese *roji* (tea gardens); an iconic twentieth-century American modernist garden; the 'international' willow-pattern design; geology and designed landscapes; gnomes in twentieth-century New Zealand; and landscape authorship of a nineteenth-century New Zealand public garden.

Each article also examines transfers of cultural ideas, and the material and power dynamics underlying them. Each article develops a particular methodology to enable examination of exchanges of complex cultural forms and their embodiment in gardens. This can reveal that seemingly straightforward pathways of exchange, such as from China to Japan in the case of tea gardens, can belie the complexity of cultural exchange and the difficulty of understanding and unravelling such transfers. On other occasions, geographical movements of ideas appear to be inextricably complex, as through the willow-pattern design's arrival in New Zealand via the Middle East, China, and Britain, yet their cultural denouement is not.

A key methodological challenge linking the issue is the extent to which scholars can write cultural histories of gardens whose physical presence has long since disappeared or whose only archive is the site of the garden itself. Another important theme considered in the issue in exchange is the role of particular sites, and their physical attributes, in moderating cultural and design intentions and uses.

Moving from text, to aesthetic practice, to physical site, in 'Chinese Sources in the Japanese Tea Garden', Richard Bullen uses literature and poetry to trace the aesthetic influence of Chinese philosophy and Chan (Zen) Buddhist practices on the development of Japan's *wabi* tea culture and, in turn, on the design and use of sixteenth- and seventeenth-century tea gardens (*roji*), key sites for the formative tea ceremony (*chanoyu*). Such a methodology is necessary as there are no surviving tea gardens from this period and no direct textual reference to Chinese–Japanese tea-garden exchanges. Instead, as Bullen argues, in encouraging a broader taste for that which is unaffected or ordinary, tea masters translated the influence of Chinese religious and literary thought on *wabi* tea culture into a preference for the vernacular in *roji*, expressed in everything from the use of only native plants and a limited colour palate, to the seemingly naturalistic placement of rocks and the simplicity of locally sourced, rustic tea implements. *Roji*, Bullen shows, embodied and reflected the culture of the *chanoyu* tea masters fully, for, as he notes, 'somewhat paradoxically … Chinese philosophy and literary theory inspired medieval Japanese tea masters to source the local and vernacular for their gardens'.

Cultural exchange and responsiveness to particular *genii locorum* is also the subject of James Beattie's article, 'China on a Plate: A Willow Pattern Garden Realized'. Using poetry, pottery shards, artistic depictions, archaeological evidence, site analysis, photographs, and newspaper accounts, Beattie examines the cultural history of willow-pattern ware — a fanciful oriental fantasy fabricated in late-eighteenth-century Britain, but which had its origins in earlier cultural contact between China's Yuan Dynasty (1271–1368) and the Muslim world. Beattie shows how the story of Hawera's Willow Pattern Garden affords a curious and fascinating example of the multiple meanings *chinoiserie* elicited — not least, its inspiration for musical productions, plays, and poetry, diplomatic put-downs, critiques of colonialism and its three-dimensional representation in garden form. A broader discussion of the willow-pattern design's meaning in New Zealand suggests the need to reconfigure spatial and cultural understandings of cultural encounters away from simply experiences of intimidation and domination, as suggested by Mary Louise Pratt.

Just as both Bullen and Beattie make a strong case for the importance of local contexts in shaping cultural ideas and garden use, so in 'On Loanwords and *Calques*: Where the Language of Design Meets the Language of

Geology', Jacky Bowring examines the appropriation of local geological features and their translation into a new design language. Like Bullen and Beattie, she makes a strong case for the value of textual evidence in garden history. But in Bowring's case, she 'proposes a model of linguistic analogy' to inform a new methodology for 'probing how garden elements are deployed when an imported model is placed within a local context'. The linguistic concepts she draws from are loanwords and *calques*. The latter refers to a tracing from one language into another, whether of a phrase or word. In the sense in which Bowring uses it, '"landscape *calques*" occur when landscape elements are translated by the colonizing design language into a mimic of their original form', as in the case of Battersea Park's faux geology of manufactured Pulhamite rock. Landscape loanwords occur through incorporation of natural features, like Christchurch Botanic Garden's remnant sand dune, into new design compositions, to the extent that while an original feature might retain its physical presence, it is now understood within a more recent design context.

In 'Gardens, History and the Designer: Contributions to Historiography', Ian Henderson also examines geological features, in addition to other aspects of the natural environment, as means of articulating the importance of considerations of site and broader landscapes to the analysis of gardens. He does so by examining what for many landscape designers is the very quintessence of modernism, El Novillero, Sonoma County, designed for the Donnell family by Thomas Church in 1949. Based on discussion of the Donnell Garden, Henderson makes an impassioned plea for garden historians to take account of the physical presence of gardens, including their sensate qualities, in any analysis of them. He coins the term '*landscapeness* to describe the materiality and tangibility of gardens — encompassing ground and vegetation and exposure to the elements, and the three-dimensional environment of gardens — being in and of a garden'.

Using his concept of *landscapeness*, Henderson's analysis of the Donnell Garden adds considerably to existing interpretations of this garden, and of modernist gardens in general. Existing scholarship overwhelmingly emphasizes the appropriation in modernist landscape architecture of wider art forms, giving little or no consideration to the responsiveness of such gardens to their broader physical setting. As Henderson notes, if in the Donnell Garden '[t]he apparent reiteration of some Jean Arp or Miró-style artwork seems a somewhat formalist object in plan form', 'when viewed from the position of sitting in or just outside the lanai, the line of the pool's edge appears to echo the forms of the waterways winding across the valley which this terrace overlooks'. Similarly, the colours of the pavers, groupings of rocks, and plantings reflect the colour and form of the wider landscape and its elements in which the Donnell Garden is situated.

Christina Dyson examines landscape designs of the same period as the Donnell Garden, but in her case, she analyses the 'natural' garden in urban Australia. In 'Rethinking Australian Natural Gardens and National Identity, 1950–1979', Dyson uses Pratt's idea to argue that 'the garden provides a space in which relationships among cultures and between cultures and place can be negotiated'. She outlines her argument by focusing on a neglected topic from a neglected period of historical inquiry — the much-maligned 1950s to 1970s, wherein a handful of mainly urban landscape designers sought new expressions of Australian nationalism through development of a vernacular garden. Dyson successfully reverses the unjustified neglect of this period and its perception as a benighted hyphen between the foundation, early expression and entrenchment of white Australian nationalism and its increasing challenge from the late 1970s.

Dyson shows how, like others before them, some post-war Australian garden writers and landscape designers sought inspiration from Australian landscapes and native plants. Yet post-war writers differed from their predecessors, she argues, in the extent to which they reimagined Australia's ancient landscape and its usefulness to Australian nationalism. Although differing in conception and interpretation, all of the post-war designers deployed elements of Australia's geological antiquity to foster, as Dyson notes, 'a shared sense of history and a national heritage and culture that stood in the place of European Australia's Anglo-European cultural inheritance'.

Dyson drew on published writings, photographs, and designs of sites long since lost or never in fact created. This contrasts with the subject of Michael Roche's paper — 'W. W. Smith and the Transformation of the Ashburton Domain 'From a Wilderness Into a Beauty Spot', 1894 to 1904'. In commencing research on Ashburton Doman, Canterbury, New Zealand, Roche found that aside from two perfunctory letters, its Curator, W. W. Smith, had left no written record of his design intentions. Instead, Roche has had to reassemble Smith's landscape authorship through postcards, maps, and general descriptions of the Domain; working, in effect, from site to paper. Through detailed site analysis, Roche reveals the cultural importance of European aesthetics in the creation and adornment of Domains — public spaces which, despite their ubiquity and importance in most New Zealand settlements, have been almost entirely neglected by scholars. Roche also discerns in Ashburton Domain the traces of Smith's subsequent promotion of indigenous plants and nationalism, which found full expression in his appointment to New Zealand's Scenery Preservation Commission and subsequent employment as Curator, Pukekura Park, New Plymouth in the twentieth century.

The final article examines a neglected, but for several decades ubiquitous, presence in gardens around the world: the garden gnome. In 'The Cultural History of the Garden Gnome in New Zealand', Ian Duggan challenges current stereotypes of the garden gnome and attempts to correct an imbalanced garden historiography that has overlooked this form of garden statuary. Duggan shows that in New Zealand garden gnomes were initially associated with elite culture and only became popularized in the 1950s, thanks to cheaper methods of production and increasing affluence among New Zealanders. While New Zealand's experience mirrored the process of the migration of garden gnomes from elite into popular gardens that occurred in Britain, Duggan shows that the timing of their introduction into New Zealand did not. In Britain, garden gnomes were first introduced in the mid-nineteenth century, gaining popularity by the late 1800s; in New Zealand, they first appeared in the 1930s, and only enjoyed widespread popularity in the post-war period.

Like Beattie's article on willow-pattern ware, Duggan's also highlights the cultural adaptability of a design concept which in recent years has come to embody everything from commercial advertising to national identity — and which has returned full circle to become once more an object of elite consumption, in the case of Gregor Kregar's public sculpture of a gnome. While both apparently tell histories of ephemeral, fanciful designs, Duggan and Beattie's works alike ask researchers the serious question of just what determines their selection of research topics, and why garden historians too often ignore non-elites. Garden historians ought to take seriously Duggan's seemingly light-hearted question about what we can learn of the cultural significance of the pink flamingo in our gardens.

Acknowledgements

The special issue draws from a symposium of the same name, held at Hamilton Gardens, New Zealand, from 29 to 31 January 2014. I thank the sponsorship of the University of Waikato and Hamilton Gardens for that event, and the support of Hamilton Gardens Director, Dr Peter Sergel, and his staff: Kylie Burness, Amanda Graham, and Ceana Priest. For the special issue, I thank the copy-editors, Sarah-Mae Berry and Dr Austin Gee. Finally, I thank the encouragement of John Dixon Hunt, and the support of all of the contributors and referees.

Disclosure statement

No potential conflict of interest was reported by the author.

Funding

A Faculty of Social Sciences Contestable Research Grant, University of Waikato, funded this project, including the travel and research assistance which made possible this special issue.

NOTES

1. Mary Louise Pratt, 'Arts of the Contact Zone', *Profession*, xci, 1991, p. 34.

2. David Porter, 'Beyond the Bounds of Truth: Cultural Translation and William Chambers's Chinese Garden', *Mosaic: A Journal for the Interdisciplinary Study of Literature*, xxxvii, 2004, p. 56.

Chinese sources in the Japanese tea garden

RICHARD BULLEN

Introduction

This paper examines the tea garden (*roji*, sometimes *chaniwa*), which evolved in the sixteenth century as an aspect of the preference for *wabi* taste, through Japanese cultural negotiations with China. The paper argues that design ideas represented in *roji* and informed by *wabi* are evidence that the taste of formative tea ceremony (*chanoyu*) masters was deeply informed by Chinese culture. Texts, such as the *Laozi* and *Analects*, were primary cultural sources for the literary elite of the Japanese medieval period, and through them, for the sixteenth- and seventeenth-century tea masters. Therefore, the paper considers *roji* as complex sites where the intellectual traditions of the ascendant culture of China, transmitted by way of religious philosophy and literary theory to Japan, were realised in the form of a garden type.

Although no sixteenth- or seventeenth-century *roji* exist today in their original state, through contemporary texts we can gauge something of their character and features. First, I examine the way in which a key principle of Zen informed tea taste in the *roji*, before turning to the Chinese literature and literary theory which had such a deep impact on Japanese literature. I show that one form of poetry in particular — *renga*, or linked verse — had an important influence on the formative years of the developing culture of *chanoyu*. By way of prelude, I introduce the term '*wabi*'.

A vast literature exists describing the work of Sen no Rikyū (1522–1591). He is generally credited with crystalizing the form of tea ceremony as it is carried out today, and has had a profound influence in the world of *chanoyu*, and beyond. As Yoshisada Ishida has observed, there is a large literature on the evolution of tea-drinking rituals in Japan up to the time of Rikyū, and the

term thought to express his taste — *wabi*. *Wabi* is a deeply uncertain and problematic term, as I demonstrate.[1] The general lack of evidence on the history of tea, including even that on the steps Rikyū took in developing the *wabi* style of tea practice (*wabicha*), led Ishida, in exasperation, to ask: 'Isn't it because tea possesses within itself a great wasteland of obscurity that prevents understanding by modern people, especially modern Japanese?'[2] A consequence of this lack of clarity at the very centre of *chanoyu* studies is that *wabi* has evolved as a highly vexed term, subject to wildly diverse claims and interpretations. A brief survey of the kinds of interpretations offered illustrates the point and justifies the later interpretation put forward by this paper of the influence of Chinese ideas on the development of the stylistic representation of *wabi* in garden form.

A popular construction in *chanoyu* research, both in Japan since the seventeenth century and the modern West, is to understand *wabi* as representing spiritual and ethical aspirations, particularly that of Zen (China; *Chan*). Paul Varley and George Elison note that 'many writers have all too facilely pictured the tea ceremony as a natural or logical manifestation — even as the inevitable distillation — of Japanese Zen'.[3] For example, the *Zencharoku* (*Zen Tea Record*), a document probably written in the nineteenth century by a Zen priest, described *wabi* as meaning that:

> [E]ven in straightened circumstances no thought of hardship arises. Even amid insufficiency, you are moved by no feeling of want. Even when faced with failure, you do not brood over injustice.[4]

Koshirō Haga explained this passage in the following way:

[W]abi means to transform material insufficiency so that one discovers in it a world of spiritual freedom unbounded by material things. It means not being trapped by worldly values but finding a transcendental serenity apart from the world.[5]

In the West, two widely read Japanese advocates of a Zen-inspired interpretation of chanoyu and wabi writing in the mid-twentieth century were Daisetsu Suzuki and Shin'ichi Hisamastu. Suzuki's definition of wabi in Zen and Japanese Culture (1959) mirrors that described by Haga: 'To be poor … and yet to feel inwardly the presence of something of the highest value, above time and social position: this is what essentially constitutes wabi.'[6] Understanding wabi in these terms consequently proved popular with Western scholars and writers. In 1992, Gerd Lester, for example, writes that it 'advocates a life of austerity, the renouncing of material attainments for spiritual enrichment'.[7] Wabi, together with the term with which it is often conjoined, sabi (as wabi-sabi), was picked up by American counter-culture in the mid-twentieth century. Leonard Koren, the author of the popular text Wabi-Sabi for Artists, Designers, Poets and Philosophers, for example, confesses:

Like many of my contemporaries, I first learned of wabi-sabi during my youthful spiritual quest in the late 1960s. At that time, the traditional culture of Japan beckoned with profound 'answers' to life's toughest questions. Wabi-sabi seemed to me a nature-based aesthetic paradigm that restored a measure of sanity and proportion to the art of living.[8]

The architectural historian, Teiji Itoh, by contrast, defines the 'core of wabi' in very different terms, as:

The refined and elegant simplicity achieved by bringing out the natural colors, forms, and textures inherent in materials such as wood, straw, bamboo, clay and stone, as well as in artifacts crafted from them … .[9]

And again, the historian Matsunosuke Nishiyama argues that wabi developed as a politically charged aesthetic counter to the ostentatious taste of military rulers:

The stress on unqualified simplicity in the [seventeenth-century] Nambōroku was a form of protest against contemporary attempts by members of the ruling elite to revive the old-style banquet style.[10]

Kendall H. Brown applies the theories of anthropologist Victor Turner in describing wabi-style chanoyu as 'an excellent model of ritual antistructure', and emphasizes the role of the roji as the passage towards fictional reclusion.[11] Hence, 'the roji designer had to deploy a variety of symbols to suggest a mountain wilderness', such as the sleeve-brushing pine (sodezuri matsu), and washbasin (tsukubai).[12]

Pertinent to this paper is Marc Keane's detailed 2009 analysis of the history of the tea garden, in which he devotes a section to wabi. In it, describing the influence of renga (described below) on chanoyu, as well as wabi's literary expressions outside renga, he concludes that wabi:

[C]an be seen as a form of beauty, a melancholy beauty such as found in autumn, golden-warm yet tinged with the cold of oncoming winter. It can be seen as a sense of loneliness as felt by a hermit who voluntarily removes himself from society, but also a bitter loneliness, the awareness of the inherent human condition — of being utterly alone with oneself.[13]

Despite being assured that the understanding of terms like wabi, 'sheds some light on the art and practice of chanoyu',[14] the rest of Keane's work is virtually silent on the subject of wabi. Nor is wabi cited to aid in the reader's comprehension of sixteenth- and seventeenth-century tea gardens' character or distinctive features.

Further, an evident difficulty of Keane's account of wabi— as, indeed, to a lesser or greater extent of the other accounts cited above — is that it is of questionable usefulness in understanding the character of objects described as expressing wabi, such as bowls, vases, buildings or gardens. How can a bowl, for example, express 'a bitter loneliness, the awareness of the inherent human condition — of being utterly alone with oneself'? Indeed, how can it manifest 'transcendental serenity apart from the world'? The answer to this conundrum, plainly, lies in the fact, as the philosopher Tetsuzō Tanikawa noted, that wabi— like many other aesthetic ideas in East Asia — 'is not simply an esthetic [in the Kantian sense], but an attitude toward life that has ethical and religious meaning'.[15] The present study, in contrast with Keane's, presents an account of a largely overlooked aspect of the wabi aesthetic, informed by Chinese culture and its manifestations in certain characteristics and features of tea gardens.

Chinese culture and tea in medieval Japan

For an extended period from the end of the twelfth century — that is, through the Kamakura (1185–1333) and Muromachi (1333–1568) periods — Japan revived contact with the powerful intellectual and cultural force of China. Zen was brought to Japan at the beginning of the Kamakura period, along with cultural forms associated with this belief-system, such as *gazen* (painting in black ink) and Zen-informed garden design. Chinese literature from all periods, and paintings, particularly from the Song (960–1279) and Yuan (1280–1368) Dynasties, were carried back by Japanese priests who travelled to China to study Zen at the old Southern Song capital of Lin'an (Hangzhou). It also arrived through trade with Ming China (1368–1644), which was re-established by the Ashikaga shoguns in the Muromachi period.

By the fifteenth century, a distinct style of painting developed in Japan which regarded Chinese paintings as 'bench marks which set the standards for the ink painters of the day'.[16] A canon of Chinese paintings was established by the third (Yoshimitsu, r. 1368–1394), sixth (Yoshinori, r. 1428–1441) and eighth (Yoshimasa, r. 1449–1473) Ashikaga shoguns, their aesthetic advisors, and influential Zen temples. A great number of works by artists favoured by the Japanese were brought to Japan, some of which became far better-known and appreciated there than in China. A catalogue of paintings held in the eighth shogun's collection, for example, lists 107 paintings by the late Song/Yuan Dynasty *Chan* monk Muqi (n.d.), whose works were amongst the most highly prized by the Japanese, although he remained largely unrecognized in China. Likewise, paintings by artists such Yujian and Liang Kai (both active in the thirteenth century) became highly sought after in Japan, but were of minor status in China.[17] Medieval Japanese painters also modelled their work on these Chinese paintings, copying their brush technique and subjects, as well as the overall ambience of the original artist's work.[18] The structure and aesthetic of the tea ceremony developed in this world — a world in which Chinese cultural achievements and imported Chinese artefacts set the standards for artistic accomplishment.[19]

The 'golden age' of *chanoyu* was the period of the great *chanoyu* masters, such as Sen no Rikyū, Furuta Oribe (1544–1615) and Kobori Enshū (1579–1647). Simply put, at a tea ceremony as it was practised by these men and is still practised today, a host receives and entertains a small number of guests with a meal, *sake*, and tea, which he prepares in front of them. This typically takes place in a small building — a teahouse (*chaya, sukiya*) — set within a *roji*.

The *roji* itself is separated from the outside world by a fence. In contrast to other Japanese garden types, borrowed scenery (*shakkei*) does not ordinarily form part of its composition: *roji* tend, then, to be very inward-focused and sealed environments. Although sizes of *roji* vary greatly, the *roji* — standardized in the Edo period (1603–1867) — allows for a small waiting house (*machiai*), an 'outer garden' (*soto niwa*) containing an outside waiting arbour (*koshikake machiai*), a fence with a 'middle-gate' (*chūmon*), and an 'inner garden' (*uchi niwa*), which contains a water-basin and teahouse. A pathway leads from the *machiai*, through the *chūmon*, to the entrance of the teahouse.

As the tea ceremony evolved it became governed by complex rules, which applied to every aspect of the procedure, including the environment (the teahouse and garden) in which it was practised. Sōshitsu Sen writes:

> In referring to the rules for the conduct of the tea service, it is customary in the Way of Tea to use the phrase '*kiku sahō*', literally, 'standard etiquette'. These rules govern not only the design of the tearoom and the adornment of its alcove, but also the garden outside. In this sense one can say that everything about tea practice follows these rules.[20]

Rules of tea and conventions which apply inside the teahouse and to the actions of practitioners, provide background to understanding both the design principles of the *roji* and the impact of Zen culture in the developing aesthetic and design of these gardens.

Frugality of Zen in tea culture and gardens

Zen Buddhism arose in China in the Tang Dynasty (618–907), was introduced into Japan at the end of the Heian period (794–1185), and became the leading Buddhist sect amongst the ruling military classes in the succeeding Kamakura and Muromachi periods. Zen played a key role in the history of tea in Japan: both the tea plant (*Camellia sinensis*) and the method for making the powdered tea drink, were introduced as part of Zen culture. In the thirteenth century, tea cultivation and drinking was largely confined to Zen temples. Over subsequent centuries, ties between tea and Zen remained close, as captured in the well-known saying *chazen ichimi*, 'the taste of tea and Zen are one'. Most of the formative sixteenth- and seventeenth-century tea masters, such as Take no Jōō (1502–1555), Rikyū, and Enshū, were trained in Zen temples.

Zen, which is directed towards the inner experience of enlightenment, considers objects in the outer world — especially those which are attractive or colourful — as distracting. This accounts for the often austere character of Zen temples, and their gardens, and the preference in Zen culture for monochrome ink painting. Rejection of the sensuously pleasing also informed *chanoyu* practice. The value of austerity is spelled out clearly in texts on *chanoyu*, especially those inflected with Zen understanding, for example the *Zencharoku*, cited above:

> To crave exotic goods and rare treasures, or pass judgement on the delicacy of wine and food, or build for entertainment, making your tearoom resplendent and toying with trees and stones of the garden, is to be at variance with the original significance of the Way of Tea.[21]

The drive towards frugality is expressed in the architecture of tea. 'In appearance', as Shuichi Katō wrote, 'the tea ceremony room resembles a small country cottage; any appearance of lavishness — comfort, even — is deliberately shunned.'[22] Typically, wooden members are left unpainted, and walls are made from clay, with a grey or rust colour added. Windows are deliberately small, and teahouses have wide overhanging eaves, so even in the daytime the light in the room is dim. One of the standard times for a tea gathering — of which there are seven — is beginning in pitch-black at 4 a.m. (artificial illumination is not used inside the room).

Deliberate starkness and deprivation is expressed in the *roji*, marking it as very different to many other Japanese garden-design types, such as the Heian-period *shinzen* court garden or Edo-period stroll garden, in which the spectacular colours of blossoms or autumn leaves are important components.[23] In the *roji*, bright colours, such as flowering trees, are shunned, as are curious or interestingly shaped rocks or shrubs. As the twentieth-century practitioner, Sen'ō Tanaka, explained: 'Flowering trees are avoided on purpose, for it was said that if one's eye were distracted by the *roji*, one's mind might be distracted from the mood of the tea ceremony'.[24]

According to historical records, some sixteenth-century gardens were particularly sparse. The *Sukidōshidai* (*Senrin*), written in the sixteenth century, claims that the *roji* should not contain *any* plants or stones. A garden of which we have a description, designed by Hisamasa (d. 1598) in 1587, was

FIGURE 1. *The strikingly empty garden (*roji*) of the* Meimei-an *teahouse, built in 1799 in Matsue, as it is today. Source: photograph by author.*

of this character, comprising moss and just one tree, and according to Kinsaku Nakane: 'One of Rikyu's *roji* was so simple it did not even have stepping stones.'[25] The *roji* of the *Meimei-an* teahouse (built in 1799) in Matsue is just such a strikingly empty garden that can be visited today (figure 1). The *Meimei-an*'s builder, Matsudaira Fumai (1751–1818), cited the following poem in his directions to the Tea practitioner, urging a favouring for the stark and plain:

> When the land is bare
> And the fields are colourless
> Then the view is best.
> Better than autumn tints
> Is this lack of colour.

According to the *Nampōroku*, Take no Jōō chose the following verse by Fujiwara no Teika (1162–1241), which expresses the same distaste for the colourful sights conventionally enjoyed:

Looking about
Neither flowers
Nor scarlet leaves,
A bayside reed hovel
In the autumn dusk.[26]

Both of these oft-cited verses are well-known to *chanoyu* practitioners. A rule applied to *roji*, quoted by Shōzō Okuda, runs: 'For purification of the mind, not the delight of the eye'.[27] The deliberate deprivation of sensory pleasure taught by Zen formed a core component of the design of the *roji*. To this Zen-sourced-quality of *chanoyu* practice are added further cultural influences, to which we now turn.

Ordinariness or blandness in Chinese and Japanese literary and art theory

One of the sources for the early *chanoyu* masters for the convention of removing themselves from the main residence in order to pursue a cultural or aesthetic activity, came from the literary cultural activity that dominated Muromachi-period Japan, known as *renga*, or linked verse poetry. At *renga* parties, groups of friends would gather, at times in purpose-built structures set in gardens separated from the main residence, or in natural settings, and compose poetic links. Participants contributed links of 5–7–5 and 7–7 syllables towards the poem. The buildings were sometimes referred to as grass-thatched cottages (*sōan*), built in the manner of mountain-village hermitages, and described as 'retreats in the midst of the city', or 'mountain hideaways' (*yamazato*). *Sōan* and *yamazato* were both terms adopted by tea masters into *chanoyu* to describe teahouses and their settings. Dennis Hirota noted that *chanoyu* and *renga* 'stand out as particularly close sibling arts' of the Muromachi period, writing that:

> Both renga and tea are characterized by minutely detailed formal conventions, which were gradually evolved in order to nurture the performance and to maintain a high esthetic standard. In these, renga was the forerunner, and probably many of its discoveries were passed on into tea.[28]

The formative years of the tea ceremony coincided with the height of *renga* culture, and the poets recognized as the finest exponents of the art, Shinkei

(1406–1475) and Sōgi (1421–1502), were contemporaries of Murata Jukō (1423–1502), the person attributed in conventional *chanoyu* histories to have set *chanoyu* on the course towards what would be the *wabi* style of tea. Jukō and Take no Jōō, another leading figure in the history of *chanoyu*— and Rikyū's teacher — knew the foremost men in the world of *renga*, read *renga* theory, and themselves joined in *renga* parties, sometimes serving tea at the same gatherings. Beyond meeting with a small number of people in a space separated from the mundane world in order to share in a cultural pursuit and in addition to the developing overwhelming presence of rules, it is clear from numerous historical sources that practitioners of tea aimed to adapt into *chanoyu* aspects of *renga*'s aesthetic. For example, Jukō, in a letter known as *Kokoro no fumi*, makes direct reference to an aesthetic taste expressed in *renga*, the 'cold and withered' (*hieyase*); and, according to a passage in the diary of Yamanoue Sōji (1544–1590), Jōō, too, cited Shinkei's aesthetic preferences for *hieyase*.[29]

The work of Shinkei expressed a preference for the ordinary, plain or bland. For example, in *Hitorigoto*, he wrote: 'The finest *renga* are like drinking water. They have no special flavor, but one never tires of them, however often one hears them. Unusual things are interesting when one happens to hear them, but gradually they lose their interest.'[30] His instruction manual on *renga*, *Sasamegoto*, likewise, contains passages, such as the following, which recommend plainness of style:

> In true ears, the verse of poets who deck out their feelings has the ring of deception, however graceful in form and diction it may be. Among the famous poems of old masters, and those of their works which they themselves approved, a decorated surface is rarely to be found. In particular, the poems of the ancient period are so straightforward and direct that the eyes of this decadent world, accustomed as they are to affectation, cannot discern their excellence.[31]

In the same manual, Shinkei wrote: 'As a rule, a style that stresses simplicity and ease can be said to manifest the true way.'[32]

In turn, the source of Japanese literary figures' preference for the simple, ordinary, or unaffected, was the artistic theory and philosophy which came to Japan from China from the thirteenth century. Chinese philosophical texts, widely read in Japan by educated men, played a central role in forming medieval taste. They came from across the intellectual spectrum, including Confucianism, Daoism and Buddhism, as well as literati culture. In addition to the value of frugality reinforced by these works, ordinariness, blandness, or lack of affectation, was praised. The *dao* itself was said to have no flavour. Laozi taught:

> Music and the dainties will make the passing guest stop (for a time). But though the Tâo [*dao*] as it comes from the mouth, seems insipid and has no flavour, though it seems not worth being looked at or listened to, the use of it is inexhaustible.[33]

Blandness was appreciated initially as a personal quality, a favourable characteristic of men. As Zhuangzi wrote: 'Dealings with the gentleman are as bland as water, while dealings with the small man are as pleasing to the taste as new wine.'[34] Confucius, likewise, made clear his favouring of behaviour that is unaffected. In the first book of the *Analects*, it is recorded: 'Clever talk and affected manners are seldom signs of goodness.'[35]

In the literature of the Song Dynasty, the quality of blandness became a guiding aesthetic principle.[36] Texts of Song Dynasty poetry theory entered medieval Japan, and formed part of the cultural learning of the educated.[37] An example is the poetry-writing manual *Shiren yuxie* (1244) by Wei Qingzhi (fl. thirteenth century), which quoted authorities to highlight the value of plainness and ordinariness, and to criticize that which was skilful or florid.[38] More generally, Confucian and Daoist texts were widely read in Japan from the time that government based on the Chinese model was established in the Nara period (710–794), when literacy became the norm amongst government officials and priests. There is a charming section in the *Tsurezuregusa*, written in 1333 by the retired priest Yoshida Kenkō (1283–1350), whose work exerted an enormous influence on later medieval Japanese intellectuals, in which he advised reading Chinese classics:

> The pleasantest of all diversions is to sit alone under the lamp, a book spread out before you, and to make friends with people of a distant past you have never known. The books I would choose are the moving volumes of *Wen-hsüan* [*Wen Xuan*], the collected works of Po Chu-I [Bai Juyi], the sayings of Lao Tzu [Laozi], and the chapters of the *Chuang Tzu* [*Zhuangzi*].[39]

In various sections of the *Tsurezuregusa*, Kenkō represents his own taste as aligning with that of Chinese authorities, and criticizes those who exhibit a garish and unrestrained taste. In §139, he turns his attention to the plants he would like most in his own garden: cherries, pines, plums, willows, and the like. And he adds:

> Ivy, arrowroot vine, and morning glories are all best when they grow on a low fence, not too high or too profusely. It is hard to feel affection for other plants — those rarely encountered, or which have unpleasant-sounding Chinese names, or which look peculiar.
>
> …
>
> As a rule, oddities and rarities are enjoyed by persons of no breeding. It is best to be without them.[40]

In his garden, Kenkō preferred plants which were native, or common to Japan, over exotic or unusual plants.

Texts such as the *Laozi, Analects*, and Kenko's *Tsurezurgusa*, were primary cultural sources for the literary elite of the Japanese medieval period, and through them, for the sixteenth- and seventeenth-century tea masters.

The tea garden

The aesthetic principles of ordinariness and plainness permeated every aspect of the tea ceremony that developed in line with *wabi* principles: from the garden, teahouse, scroll, and flower arrangement decorating the room, down to even the way of preparing tea itself. To the uninitiated, the methods of preparing tea seem long and complex. In fact, they were designed and refined by tea masters over centuries to embody the most simple and efficient actions for making and offering a bowl of tea to guests. The aim of the host is not to impress in the manner of a conjurer, and although he is trained and the actions are highly self-conscious, the practitioner aims to carry out the ceremony as if it were not a series of deliberate acts of rule-following. Two of Sen no Rikyū's so-called 'Thirty-five Dislikes in the Way of Tea' (*Rikyū sanjūgo-jō kenki*) affirm these ideas. First, *yu-jaku* (literally, oil ladle), admonishes the practitioner for flamboyant actions: 'When pouring water from the ladle don't lift it up imitating the showy action of an oil salesman demonstrating the consistency of his oil.' The second, *nemura* (literally, sleepy,) warns against carrying out tea procedure with a mind only to following the rules: 'If you become too concerned about each step of the procedure your *temae* will lack life and it will seem as though you are doing it in your sleep.'[41] Both of these purportedly accord with Rikyū's own practice, of which it was written:

Rikyu's way of serving tea was not elaborate at all. Neither at its start nor at its finish could any conspicuous point of tastefulness be recognised. This can be called the very essence of tea, far from simple normality or routine commonness.[42]

These values of ordinariness, plainness, and the apparently artless, were carried into *roji*, which in the style associated with the *wabi* aesthetic are usually deprived of anything that stands out or might catch attention (figure 2). The most cited rule regarding the *roji* is that of Rikyū, who instructed that it should resemble a little-used mountain path.[43] Okuda wrote the following: 'It is said that the roji embodies the atmosphere of a path through a meadow. Avoid artifice. Even in the planting of trees, conspicuousness is to be rejected.'

In another passage, he wrote:

> Rare and unusual forms are avoided, and emphasis is placed on the ordinary and unpretentious. Thus, it is sufficient to compose the atmosphere through the arrangement of plants and trees that are native to the mountains and meadows, together with common rocks. It is, of course, unacceptable to seek out the quaint and curious, but it is also undesirable to take special pains to find plain things.[44]

This passage alerts us to the idea that the common or ordinary is that which is to hand, is native, or easily accessed. The local or vernacular is favoured over that which is rare or difficult to attain. Kenkō cited Chinese sources to make the same point: 'I believe it is written in the classics somewhere, "He did not prize things from afar", and again, "He did not value treasures that were hard to obtain."'[45]

According to Matsudaira Fumai, the *roji* of the *Yū-in* teahouse at the Konnichi-an in Kyoto was 'the best model'.[46] The garden is typical, in that colours other than the green of the foliage, and brown and grey of the ground are absent, and there is nothing unusual that will attract the eye. Typical of *roji*, the plants are common in Japan. The demand for that which is inconspicuous or ordinary has a parallel in a rule Rikyū gave for the flower arrangement that decorates the tearoom; he instructed: 'Arrange the flowers with the feet.' By this, Rikyū is taken to mean that only plants which one may gather oneself, that is, that grow locally and are in season, are to be used. Unusual or rare flowers, or those blooming out of season, are considered incorrect. The same prohibitions, as Okuda noted above, apply to the *roji*. Another of Rikyū's rules

FIGURE 2. *The Tai-an roji, in the style associated with the* wabi *aesthetic, is deprived of anything that stands out or might catch attention and thus displays the values of ordinariness, plainness, and the apparently artless. Source: photograph by author.*

for the flower arrangement — '[a]rrange the flowers as they are in the field' — directs the tea practitioner to create an arrangement that, paradoxically, appears unarranged. Again, this is a parallel inside the teahouse and the desire in the *roji* for the quality of ordinariness, or 'artlessness', so as for it to appear as if it came about naturally, without human intervention (figure 3).

The laying of the stones in the *roji* is a matter of considerable interest for *chanoyu* masters, the rules for which instantiate these ideas. Okuda noted ' … the stones should be just large enough to accommodate the foot, and a fine laying [sic] of a poor stone reveals the skill of the host'. He continues: '"Fine" here does not mean with obtrusive ingenuity'. According to Rikyū's grandson Sōtan (1578–1658), however, the placement of some of the stones could be set by chance:

> Rikyū said, concerning the rocks placed in front of the stone basin [to cover the drain], that one should have a servant close his eyes and drop them from a container; then, after arranging them slightly with a cane, one should leave them as they are. It is inappropriate to drop them deliberately.[47]

Setting the stones in this way, we may surmise, was seen to be a solution to the problem of how to avoid the unwanted quality of artifice, or 'artfulness'. The rocks should be set as if naturally occurring, and thus not attract attention. Regarding this quality, there is an instructive story of the *chanoyu* master, Kuwayama Sakon (Sōsen, 1560–1632). He was hosting a *chanoyu* gathering, and during an interval one of his guests complimented him on the layout of his rocks in the garden. After the guests had departed, Sōsen went out into his garden and rearranged them.[48]

Contemporary accounts of Momoyama-period (1568–1600) tea ceremonies and their settings are in line with this taste for the unaffected or bland, which is identified with the aesthetic of *wabi*. João Rodrigues (c. 1560 – c. 1633) lived in Japan from 1580–1610, and described the environment for tea:

> This is a special building, with a path or entrance leading to it and with various other things suitable to the purpose of the custom. In general this purpose is the quiet and restful contemplation of the things of nature … . Thus fashionable, finely wrought, and elegant things, such as those belonging to the court and not to mountains and the wilderness, are most unsuitable for this building, the eating and drinking, utensils, and everything else.[49]

FIGURE 3. *The* roji *displays the quality of ordinariness, or 'artlessness', so that it appear as if it came about naturally, without human intervention. Source: photograph by author.*

Rodrigues also described the *roji* beyond the *koshikake machiai*:

> Further on there is a wood of trees, partly natural if they were already growing there, partly transplanted thither with great skill. They choose trees that have the best shapes and branches, and the most natural and elegant artlessness. Such trees are mainly pines interspersed with others in such a way that there does not seem to be any artificiality about them, and they appear to have grown there naturally and haphazardly.[50]

In like manner to the attention the practitioner gives to his way of making tea, the *roji* is designed and assiduously tended, and thus the quality of artlessness — the end in both cases — does not come about spontaneously, but is 'achieved' (figure 4). There is an anecdote told by Sōtan which illustrates this point well:

> [During the tenth month] Rikyū and [his son Dōan] attended a morning gathering hosted by a certain person. In a morning storm, the leaves of an oak tree (*muku*) had fallen and scattered [onto the stepping stones], and the surface of the *roji* gave precisely the feeling of a mountain forest. Rikyū, looking back [on the *roji*], said, 'All of this is engaging. But the host, being unaccomplished, will probably sweep up the leaves'.[51]

The story continues that, as predicted, after the gathering's intermission, all the leaves were raked up, to Rikyū's displeasure.

A device Rikyū used to create a sense of the ordinary in the tea environment was to adopt into it everyday, mundane objects. For example, Rikyū introduced a vase into *chanoyu* made from a simple section of bamboo, based on the custom of peasants who cut a section of bamboo for carrying flowers.[52] Previously, imported Chinese celadon and brass vases were favoured by tea masters. Introducing unrefined or mundane objects into the world of tea had been undertaken by practitioners prior to Rikyū. For example, the Chinese tea jar, 'Chigusa', began life probably in the Song Dynasty as a mass-produced container for commercial products, but in Japan in the fifteenth and sixteenth centuries became one of the most celebrated *chanoyu* utensils.[53] In the *roji*, he carried out a similar creative act:

> Rikyu was attracted by the simplicity of bamboo waterpipes often seen in front of woodcutters' cottages in the mountains. These pipes lead water from

FIGURE 4. *The* Korin-in roji *is designed and assiduously tended so that the quality of artlessness does not come about spontaneously, but is 'achieved'. Source: photograph by author.*

mountain streams to old stone handmills or pails serving as wash basins. He reproduced this water system in the tea garden, for he recognized in this simple mountain waterpipe a simplicity and naturalness in accord with the principles of Chanoyu.[54]

Keane describes another way in which the ordinary was incorporated into the *roji*, and which became a tradition during the Edo period: the aesthetic device of *mitate*, the 'technique of reusing old, sometimes discarded, materials in tea gardens or as utensils in a tea gathering'.[55] These objects were called *mitate-mono*, or 'reused things'. Examples of *mitate-mono* were water basins, where guests ritually purify their lips and hands before entering the teahouse. These might be 'found' stones with a natural hollow deep enough to act as a basin, or a stone which has been used for some utilitarian or architectural purpose, in which a basin is carved. Old roof tiles also were recorded as reused in front of the basin.

Conclusion

The design principles which informed *roji*, conceived and built by six-teenth- and seventeenth-century *chanoyu* masters, were the result of Japa-nese cultural interactions with China in the medieval period. Coupled with the Zen principle of frugality, which expounded starkness and simplicity, and forbade bright colours, *roji* designs were informed by Chinese philo-sophy and literary theory, in the preference for that which is plain, ordin-ary, or unaffected. Design principles used in *roji* suggest that the preference for Japanese- over Chinese-made objects, which forms an aspect of the taste for *wabi*, can be understood as part of a broader taste for that which is ordinary (in a place), that is, a preference for the local or vernacular. The stones, plants, and their setting in *roji* suggest an intention to create an atmosphere of unaffected naturalness and ordinariness. The introduction of common or plebeian items into the tearoom was mirrored in the *roji* by the use of 'found' objects, as well as devices used by the common people. By this means, somewhat paradoxically, then, Chinese philosophy and literary theory inspired medieval Japanese tea masters to source the local and vernacular for their gardens.

Acknowledgements

My thanks to James Beattie, Rachel Payne and Douglas Horrell for their assistance in preparing this paper.

Disclosure statement

No potential conflict of interest was reported by the author.

NOTES

1. Yoshisada Ishida, *Chūsei Sōan no Bungaku* (Tokyo: Kitazawa Tosho Shuppansha, 1970), pp. 159ff. The word *wabi* was used in the Heian period (794–1185) to mean miserable or wretched. See discussion in Tetsuzō Tanikawa, *Cha no bigaku* (Tokyo: Tankosha, 1977), pp. 158–166. Translation provided in 'The Esthetics of Chanoyu Part 4', *Chanoyu Quarterly*, xxvii, 1981, pp. 41–44. Probably the best intro-duction to *wabi* in English remains Kōshirō Haga, 'The *Wabi* Aesthetic Through the Ages', in Nancy G. Hume (ed.), *Japanese Aesthetics and Culture: A Reader* (Albany: State University of New York Press, 1995), pp. 245–278.

2. The original reads: '*Sore wa omou ni, cha ga sore jishin no naka ni, kindaijin no — tokuni nihon no kindaijin no — ninshiki o kobamu you na, ōki na fumei no kōchi o motteiru kara dewa nakarou ka*'. Ishida, *Chūsei Sōan no Bungaku*, p. 159.

3. H. Paul Varley & George Elison, 'The Culture of Tea: From Its Origins to Sen no Rikyū' in Elison & Bardwell L. Smith, *Warlords, Artists, and Commoners: Japan in the Sixteenth Century* (Honolulu: University of Hawaii Press, 1987), p. 206.

4. Dennis Hirota, 'The Practice of Tea 5 — The *Zen Tea Record*: A Statement of Chanoyu as Buddhist Practice', *Chanoyu Quarterly*, xxxxv, 1988, p. 50.

5. Haga, 'The *Wabi* Aesthetic Through the Ages', p. 246.

6. Daisetsu Suzuki, *Zen and Japanese Culture* (New Jersey: Princeton University Press, 1973), p. 22. The 'san-Senke' Tea schools, Japan's three largest schools, have a wide sphere of influence, and have actively promoted this understanding of *wabi*. The twentieth-century grand Tea master of the Urasenke school, Sen Soshitsu XV (Genshitsu, b. 1923), an indefatigable advocate of *chanoyu* internationally, stresses the ethical and spiritual understanding of *wabi*, noting that it 'gradually came to take on a positive meaning, to be recognized for its profound, religious sense'. Sōshitsu Sen, 'Afterword' in Kakuzō Okakura, *The Book of Tea* (Tokyo, New York & London: Kodansha International, 1991), p. 156. Sen characterizes *wabi* as a 'spiritual aesthetic'. Sōshitsu Sen, *The Japanese Way of Tea: From Its Origins in China to Sen Rikyū* (Honolulu: University of Hawai'i Press, 1998), p. 111.

7. Gerd Lester, 'Zen Philosophy in the Japanese Tea Ceremony', *Arts of Asia*, xxii/1, 1992, p. 80.

8. Leonard Koren, *Wabi-Sabi for Artists, Designers, Poets & Philosophers* (Berkeley: Stone Bridge Press, 1994), p. 9.

9. Teiji Itoh, *Wabi Sabi Suki: The Essence of Japanese Beauty* (Hiroshima: Mazda Motor Corporation, 1993), p. 7.

10. Matsunosuke Nishiyama, *Edo Culture: Daily Life and Diversions in Urban Japan, 1600–1868* (Honolulu: University of Hawai'i Press, 1997), pp. 157–158.

11. Kendall H. Brown, *The Politics of Reclusion: Painting and Power in Momoyama Japan* (Honolulu: University of Hawai'i Press, 1997), p. 63.

12. Ibid., p. 65. Brown's theory, interesting as it is, does not address or explain the design characteristics of *roji* identified as *wabi* — plainness and ordinariness — described here. A further difficulty with Brown's theory is that it ignores the central place of rules in *chanoyu* practice, which, as I have argued elsewhere, undermine ordinary utilitarian values. See Richard Bullen, 'Freedom and Restraint in the World of Tea' in Giusi Tamburello (ed.), *Concepts and Categories of Emotion in East Asia* (Rome: Carocci editore, 2012), pp. 252–265.

13. Marc Peter Keane, *The Japanese Tea Garden* (Berkeley: Stone Bridge Press, 2009), p. 40.

14. Ibid., p. 39.

15. Tetsuzo Tanikawa, *Cha no Bigaku* (Tokyo: Tankosha, 1977), p. 165; translation from 'The Esthetics of Chanoyu, Part 4', *Chanoyu Quarterly*, xxvii, 1981, p. 44.

16. Shimao Arata, 'The Stewards of Art in Muromachi Japan: Nōami, Geiami, and Sōami', *Chanoyu Quarterly*, lxxiv, 1997, p. 12.

17. Others whose works were imported into Japan and prized there remained recognized major artists in China, such as Ma Yuan (ac. pre-1189 to post-1225) and Xia Gui (active, thirteenth century).

18. Gail Capitol Weigl, 'The Reception of Chinese Painting Models in Muromachi Japan', *Monumenta Nipponica*, xxxv/3, 1980, p. 260.

19. See Stephen Little (ed.), *Chinese Paintings from Japanese Collections* (California: Los Angeles County Museum of Art and DelMonico Books, 2014). Tea was drunk in Japan prior to the twelfth century, but the style of tea — known as 'brick tea' was different. Brick tea is still manufactured in China today.

20. Sōshitsu Sen, *The Japanese Way of Tea: From Its Origins in China to Sen Rikyū* (Honolulu: University of Hawai'i Press, 1998), p. 180.

21. Dennis Hirota (trans.), 'The Practice of Tea 5 — The *Zen Tea Record*: A Statement of Chanoyu as Buddhist Practice', *Chanoyu Quarterly*, liv, 1998, p. 41. The term *roji* itself has a foreign source. It derives from a parable in the Lotus Sutra, where a father (Buddha) saves children (ignorant people) from a burning house (corrupt world) by leading them to an open ground (*roji*). The *roji* therefore is a space free from the dust of worldly troubles and desires.

22. Shūichi Katō, *Form, Style, Tradition: Reflections on Japanese Art and Society* (Tokyo, New York and San Francisco: Kodansha International Ltd, 1981), p. 157.

23. The stroll garden, or *kaiyūshiki teien*, as a garden one passes through on foot, had its origins in the *roji*. Well-known examples, such as the Katsura Rikyū in Kyoto, are in effect large tea gardens.

24. Tanaka, *The Tea Ceremony*, p. 167.

25. Nakane Kinsaku, 'The Character and Development of the Tea Garden', *Chanoyu Quarterly*, ii/1, 1971, p. 48.

26. Matsunosuke Nishimura (ed.), *Nampōroku* (Tokyo: Iwanami Shoten, 1998), 'Oboegaki', p. 24, §33; translated and discussed in Dennis Hirota, 'The Practice of Tea 3 — Memoranda of the Words of Rikyū: *Nampōroku* Book 1', *Chanoyu Quarterly*, xxv, 1980, pp. 41–42. The *Nampōroku* may have been composed in the seventeenth century.

27. Cited in Shōzō Okuda, *Chami* (Tokyo: Kamakura Shobō, 1955), p. 38. Original reads: '*kokoro o sumasu ni arite me o tanoshimasu ni kanazu*'. Translation by Dennis Hirota from 'The Taste of Tea: Excerpts from the *Chami*', *Chanoyu Quarterly*, lxiv, 1990, p. 18. Okuda's *Chami*, originally published in 1920, was probably the most influential text on *chanoyu* practice in the twentieth century.

28. Dennis Hirota, 'In Practice of the Way: *Sasamegoto*, An Instruction Book in Linked Verse', *Chanoyu Quarterly*, xxix, 1977, pp. 26, 24. Tea was served at poetry gatherings by the early fourteenth century; see Takeuchi Jin'ichi, 'When *Karamono* Became Tea Utensils', in Louise Allison Cort and Andrew M. Watsky (eds), *Chigusa and the Art of Tea* (Washington, DC: Freer Gallery of Art & Arthur M Sackler Gallery, 2014), pp. 25–31. Traces of *chanoyu*'s early connections with the world of poetry remain in today's *chanoyu* practice, for example in the convention of guests carrying a wad of paper, used as a plate for sweets.

29. See 'The Practice of Tea 1 — Heart's Mastery: the *kokoro no fumi*; The Letter of Murata Shukō to his Disciple Chōin', *Chanoyu Quarterly*, xxii, 1980, pp. 7–24.

30. Sen, *The Japanese Way of Tea*, p. 133. Sen noted, Shinkei 'exerted extraordinary influence on Jukō'.

31. Hirota, 'In Practice of the Way: *Sasamegoto*', Chapter 41, p. 39.

32. Ibid., Chapter 12, p. 32.

33. Laozi (trans. James Legge), 'Dao De Jing', in F. Max Müller (ed.), *The Sacred Books of the East: Volume 39* (Oxford: Clarendon Press, 1891), p. 77.

34. Cited in François Jullian (trans. Paula M. Varsano), *In Praise of Blandness* (New York: Zone Books, 2004), p. 55. See Jullian for discussion of quality of *bingdan* (blandness) in Chinese cultural history.

35. Simon Leys (trans.), *The Analects of Confucius* (New York and London: W. W. Norton Press, 1997), p. 3.

36. Jin'ichi Konishi wrote that 'when the [Song] is viewed as a whole, the poets who advocated the so-called simple and crude were clearly pre-eminent, and their particular style gives the poetry of the [Song] its distinctive character'; *A History of Japanese Literature: Volume Three The High Middle Ages* (Princeton, NJ: Princeton University Press, 1991), p. 365.

37. See Konishi, *A History of Japanese Literature*, pp. 363–384, Chapter 13 'The Revival of Poetry and Prose in Chinese'.

38. See discussion in Haga, 'The *Wabi* Aesthetic through the Ages', pp. 245, 267–270.

39. Donald Keene (trans.), *Essays in Idleness: The Tsurezuregusa of Kenkō* (New York: Columbia University Press, 1967), p. 12, §13.

40. Keene (trans.), *Essays in Idleness*, pp. 123–126, §139. Kenkō's dislike for Chinese-derived names does not suggest a contempt for Chinese culture, rather for plants which are not common, or natural to the surroundings.

41. Sasaki Sanmi, *Chado The Way of Tea: A Japanese Tea Master's Almanac* (Boston, Vermont and Tokyo: Tuttle Publishing, 2002), p. 659.

42. Shigenori Chikamatsu, *Stories from a Tearoom Window* (Vermont and Tokyo: Charles E. Tuttle Co., 1982), p. 89.

43. Sōan recounts the following: 'Kuwayama Sakon asked Rikyū about making a *roji*. Rikyū said he should come to an understanding with an old poem [by Saigyō]: Yellow leaves of oak/fall and accumulate/without concern/in the loneliness of a temple road/deep in the mountains (*kasha no ha no/momiji nukarani/chiri-tsumoru/okuyama-dera no/michi no sabishisa*)'; 'Pointing to the Moon: Sōtan's Anecdotes of Rikyū's Tea', in Dennis Hirota (ed.), *Wind in the Pines: Classic Writings of the Way of Tea as a Buddhist Path* (Fremont, CA: Asian Humanities Press, 1995), pp. 254–255.

44. Okuda, *Chami*, p. 21. For translation, see Hirota, 'The Taste of Tea', p. 18.

45. Keene (trans.), *Essays in Idleness*, p. 101, §120. According to Keene, Kenkō is referring to the *Shu Jing* and *Dao de Jing*.

46. 'The Tea Maxims of Matsudaira Fumai', cited in A. L. Sadler, *Chanoyu: The Japanese Tea Ceremony* (Rutland & Tokyo: Charles E. Tuttle Co., 1998), p. 88.

47. Hirota, *Wind in the Pines*, p. 251.

48. Story cited in Tanikawa, *Cha no bigaku*, p. 95. Tanikawa wrote (pp. 94–95): 'the garden should not assert itself too boldly. As the guests approach the tearoom, should they become fascinated by the garden, its role is compromised. Beautifully manicured trees with impressive branches are unsuitable for such a path'. Translation from 'The Esthetics of Chanoyu, Part 3', *Chanoyu Quarterly*, xxvi, 1981, p. 33.

49. Michael Cooper (ed.), *João Rodrigues's Account of Sixteenth-century Japan* (London: The Hakluyt Society, 2001), p. 155.

50. Cooper, *João Rodrigues's Account of Sixteenth-century Japan*, p. 156.

51. Hirota, *Wind in the Pines*, pp. 250–251.

52. Seizō Hayashiya, 'Koyū no bunka — wabicha no sekai', in *Sen Rikyū ten* (Kyoto: National Museum, Kyoto, 1990), p. 307.

53. See Cort and Watsky, *Chigusa and the Art of Tea*.

54. Nakane, 'The Character and Development of the Tea Garden', p. 46.

55. See discussion of *mitate* in Keane, *The Japanese Tea Garden*, p. 91.

China on a plate: a willow pattern garden realized

JAMES BEATTIE

Introduction

> Like Chinese whispers, Chinese stories shift over time as they get passed on.[1]
>
> Kerry Ann Lee, *Home Made*, 2008

One of *chinoiserie's* 'defining characteristic[s]', observes David Porter, is its 'boundless adaptability, its repudiation of any fixed standard or accepted model'.[2] Those qualities are evident in the story of the development of the 'Willow Pattern Garden' in 1960s Hawera, New Zealand. It is a story that is inextricably bound up with the British culture that made the garden's motifs legible and which brought willow-pattern ware into the colony in the nineteenth century.

In detailing that history, this essay explores new geographical and methodological frontiers in garden history suggested by different interpretations of Hawera's 'Willow Pattern Garden' and the design on which it was based (see figure 6 below). Opened by the Republic of China (ROC) Ambassador to New Zealand in 1968, the story of this garden affords a curious and fascinating example of the multiple meanings *chinoiserie* elicited — not least, the ongoing attraction of the willow pattern plate in inspiring everything from musical productions, plays and poetry, to diplomatic put-downs, critiques of colonialism and the pattern's three-dimensional representation in garden form in twentieth-century New Zealand. This last manifestation of *chinoiserie* provided the newly arrived Republican Chinese Ambassador with the opportunity for taking political pot shots at the People's Republic of China (PRC). For Hawera residents, the garden had an equally important diplomatic role — not of world realpolitik, but local garden one-upmanship, thanks to the garden's promotion of the town's burgeoning civic garden culture through its realization of a playful oriental fantasy.

A broader discussion of the willow-pattern design's meaning in New Zealand suggests the need to reconfigure understandings of cultural encounters — at least their spatial and cultural expressions — away from experiences simply of intimidation and domination, as suggested by Mary Louise Pratt.[3] What can *chinoiserie*, it asks, reveal as a repository of creative encounter in New Zealand? But first to 1960s Taranaki.

The Willow Pattern Garden

On a rather blustery Saturday in early Spring 1968, dancers wearing the national costumes of 'Greece, the Ukraine, Holland, and Scotland' gathered at Hawera's King Edward Park to welcome His Excellency Mr Konsin Shah (b. 1920), ROC Ambassador to New Zealand.[4] Even if only for an afternoon, such a colourful and lively start to the weekend belied the stereotype of Hawera — and New Zealand for that matter — as a sleepy rural backwater.[5] At the time of Shah's visit, Hawera — located in Taranaki Province, on the western edge of New Zealand's North Island (see figure 9) — boasted a population of almost 8000.[6] Present-day Hawera began a little over a century earlier as a redoubt for imperial forces fighting Maori. After the land was prised from Maori and its forest removed, settlers introduced sheep and cattle. From the first decades of the twentieth century, the dairy industry has been the mainstay of the region.[7] As a rural service town, 1960s Hawera sat like an urban island of low-rise colonial bungalows amidst a sea of pea-green pastureland, on the gentle seaward-sloping Waimate Plains. Only the 55-metre-high concrete water-tower (1914) — whose neon-lit structure resembles a brightly lit squid boat — interrupted a still significantly corrugated-iron skyline.

No doubt unaware of this turbulent and still-troubled history, in his talk before a crowd of several hundred, Shah drew affinities between the design principles of the Willow Pattern Garden and the private gardens of southern China. Both, he said, had their origins in cities. Both, he noted, embodied in their designs the concepts of concealment and contrast necessary to give an illusion of great space to a small urban setting. '[B]ecause of the size of China', explained Shah, 'its people had come to enjoy small gardens in their cities. Whether large or small, the aim of a willow pattern garden was to puzzle visitors by the intricacy of design and layout.'[8] As I will show, the willow-pattern design was one born out of a creative response to the very foreignness of China — including its gardens — that the sometimes jarring and stimulating encounters with other cultures can produce.[9]

But Shah's talk ranged beyond the topic of garden design. In drawing on the standard interpretation of private Chinese gardens as places of retreat from politics, his speech took direct aim at the PRC. While 'the Chinese had become accustomed to living with and enjoying small gardens in tranquillity', he declared, ' … by sharp contrast those in Red China were unable to do this without the constant interruption of soldiers['] waving of Mao's Thoughts [sic] and the playing of martial music'. Cold War tensions were running high at the time of Shah's visit. China had embarked on the disastrous Great Proletarian Cultural Revolution in 1966, an ill-advised mass movement initiated by Mao Zedong that was marked by iconoclastic attacks on traditional culture. China was also adopting a more bellicose international stance, having in 1964 joined the questionable ranks of the atomic club.[10] An escalation of US involvement in the Vietnam War, under L. B. Johnson's administration, further heightened regional tensions. Willow-pattern ware, then, as international diplomacy?

To understand the garden's development and the cultural pull of the willow-pattern design — indeed, just how and why a fanciful design on a plate not only came to be built in three dimensions but also could embody simmering Cold War tensions — we must look to the history of the introduction of the design into Britain and thence New Zealand.

Origins

The pretty blue plate that I use at tea.
A story it has that it tells to me.
As soon as I finish my slice of bread,
A wonderful fairy tale's there instead.[11]

Queenie Scott Hopper, New Zealand, 1936

Chinese gardens were a ubiquitous presence in the nineteenth- and twentieth-century British world. From Winnipeg to Wellington, from Bombay to Ballarat, before governors and governesses, Maori chiefs and missionaries, Chinese gardens appeared at breakfast, lunch, and dinner in a clockwork regularity of culinary colonial *chinoiserie*. The blue-and-white ware tea and dinner services presented a fancifully flimsy *chinoiserie* fantasy world, 'of doll-like lovers, children, monkeys, and fishermen lolling about in gardens embraced by eternal spring'.[12] Yet, willow-pattern ware — and its distinctive depiction of a blue Chinese garden scene on a white background — became, as David Porter has written, 'paradigmatic emblems of Englishness'.[13]

'Eighteenth-century consumers in England were … infatuated with Chinese and Chinese-style goods, even as they were amused, perplexed, or troubled by the alien aesthetic sensibility these goods embodied.'[14] That visual style was, of course, not just limited to England. *Chinoiserie* came to denote 'the European manifestation of … various styles with which are mixed rococo, baroque, gothick or any other European style it was felt was suitable'.[15] Objects of *chinoiserie* designs represented, as Oliver Impey writes, 'the European idea of what oriental things were like, or ought to be like', an interpretation based on a 'conception of the Orient gathered from imported objects and travellers' tales'.[16] From its origins as a style of decorating objects such as furniture and ceramics, to the architectural layout of gardens and even buildings which were otherwise essentially European in composition, right to its later feminization, *chinoiserie* decoration 'took over the European shape and altered it more drastically' (figure 1).[17]

Perhaps the most famous manifestation of *chinoiserie* came with the development of blue-and-white ware. For world historian Robert Finlay, blue-and-white ware is the world's first global style, 'a collective visual language' in ceramic art.[18] The style developed under the Mongol Peace of the thirteenth and fourteenth centuries. Muslim merchants in Quanzhou, China, took advantage of technical improvements to Chinese porcelain enabling decorative designs in blue to be painted on highly lustrous porcelain. Until that point, Muslim artisans painted decorations on tin-glazed earthenware, but had struggled to use cobalt in decoration. The strong, yet lustrous, porcelain produced at the Imperial Workshops at Jingdezhen solved this problem. Technical improvements enabled the production of artistic designs that proved irresistible — and highly profitable — trade goods. Ships from the Middle East would bring cobalt oxide — *huihui qing* (Muslim blue) — 'over 6,000 kilometers' to China in return for 'customized wares manufactured in bulk at Jingdezhen for Islamic markets'. As a trade, it was ' … unprecedented in world history'.[19] Blue-and-white ware was eventually exported throughout South East Asia and Africa along Indian Ocean trade networks.

Relatively little Chinese porcelain reached northern Europe until the seventeenth century, whereupon the floodgates opened, thanks initially to the efforts by the Dutch East India Company (VOC). Over a 200-year period from 1700, the Dutch imported about 43 million pieces of porcelain; other countries' companies, at least 30 million.[20] European manufacturers subsequently began to produce cheap import substitutes in an attempt to cash in on the mania for Chinese objects.[21] The development of the willow-pattern ware, then, must be understood as a particular manifestation of this complex global style.

Willow pattern was the design fantasy, sometime in the 1780s, of one Thomas Minton (1765–1836), 'an apprentice engineer at the Caughley pottery'. Its full denouement came somewhat later, after Minton moved to work for Josiah Spode, an innovator who pioneered the use of underglaze transfer painting and who later perfected the recipe for finer quality bone china. Around 1790, Spode's pottery manufactory in Stoke produced the first pieces combining the elements that subsequently became known as willow pattern (figure 2).[22] Its signal features came to include 'a willow tree in the central position; three figures crossing a bridge, heading away from the main building; a zigzagging fence stretching across the foreground; and two birds hovering in the top center'. By 1814, such was the popularity of the design that many other manufacturers were copying it.[23]

The appeal of willow pattern lay not just in its attractive design; but so too in the classic tale of two star-crossed lovers with which it is associated. While there are several variations of the story, the basic plot is this: A loyal bookkeeper, Chang, works for a powerful-yet-corrupt customs official whose years

FIGURE 1. *Aside perhaps from willow-pattern ware, the Pagoda at Kew Gardens — now Royal Botanic Gardens, Kew — designed by William Chambers (1723–1796) in 1761–1762 is today perhaps the most famous manifestation of the eighteenth-century vogue for* chinoiserie, *certainly the tallest. Source: photograph by author.*

The willow-pattern ware depicts the 'chase' scene through the three figures on a bridge and their escape on a junk.

The two lovers live in a house built by Chang — their dwelling is depicted in the top-left in willow-pattern ware. Soon, however, with all of the jewellery pawned off, Chang is forced to find money through other means. He writes a book on gardens, which earns him a growing reputation, but also the attention of the mandarin. The mandarin, still seething from Chang's deceit and the loss of his daughter, orders soldiers to find his daughter and avenge her elopement. The soldiers kill Chang, while — in most stories, at least — Koong-se also dies, either by her own hand or the soldiers'. The gods pity them, and they are turned into turtle doves, which are depicted at the top of the plate.

For years, it was assumed the story came from China. Yet, in fact, it originated as a classic piece of orientalist fantasy fabricated in the factories of the British Midlands (see above); but, although fictitious, it does still illustrate both the hold of China on Western imaginations and the saturation of Chinese designs in the West. Its story and the design it depicts stood the test of time: as a New Zealand newspaper article of 1928 claimed, with some accuracy: 'The romance of two Chinese lovers, which is so quaintly pictured on willow pattern china, has greater worldwide interest attached to this design than to any other that has been evolved during the ages.'[24]

In America, John Haddad has traced the popularity of willow-pattern ware from the 1780s to the 1920s. He made the startling discovery that even several prominent, well-educated Americans actually believed the willow-pattern ware depicted the reality of China, for, as Carter Harrison opined in 1889:

Men's opinions [are] moulded, or at least colored, by the veriest trifle — coloured into prejudices, which require time and care to eradicate. He whose mother's treasured porcelain service was of the old blue willow pattern, has, more or less, his impressions of the Celestial Empire fashioned upon the model he studied upon the plates from which he ate.[25]

Haddad identifies two particular groups to whom the design especially appealed from the mid-nineteenth century. Middle- and lower-class Americans enjoyed its anti-elitism, claims Haddad, while women appreciated the female protagonist's avoidance of having to fulfil the marital wishes of her father.[26]

FIGURE 2. *Willow pattern plate. Source: From the collection of Puke Ariki, New Plymouth, A84.012. © Collection of Puke Ariki, New Plymouth. Reproduced by permission of Puke Ariki. Permission to reuse must be obtained from the rightsholder.*

of graft are about to be exposed. The mandarin resigns. So does Chang who loyally destroys his master's account books, only to be summarily dismissed. Meanwhile, Chang and the mandarin's daughter, Koong-se, have fallen in love, meeting secretly under fruit trees. Enraged upon hearing this — for he has plans to marry Koong-se to a wealthy and elderly duke — the mandarin banishes Chang from the house. But through a ruse, Chang returns and escapes with his lover on a junk, taking some of the mandarin's treasures with him too.

Along with depictions of Asiatic pheasants on plates, the willow-pattern design 'is probably the most commonly recorded pattern from archaeological sites in New Zealand'.[27] Pottery shards of blue-and-white ware found in southern New Zealand indicate its use right from the first periods of contact between European whalers and Maori in the early nineteenth century.[28] Willow-pattern ware made up some of the earliest imports into New Zealand soon after colonization in 1840, and once regular shipping networks were established, it was sold in New Zealand during the nineteenth and twentieth centuries.[29] By and large, most willow-pattern ware in New Zealand was produced by British manufacturers, rather than in China.[30] Typifying the kind of pottery popular in New Zealand is that found in New Plymouth, Taranaki's largest settlement, located some 70 kilometres from Hawera. An archaeological dig in the city unearthed willow-pattern dinnerware and other pottery, including this soup plate (figure 3), on the site of a homestead burned during the First Taranaki War in 1860. Archaeologists believe that its discovery indicates the popularity of the design among the labouring classes.[31] This is hardly surprising since from 1814 it was — and remained into that century — the cheapest pattern-print available to customers.

Indeed, willow pattern was woven into the very fabric of European culture in New Zealand, appearing on wallpaper, curtains, and becoming as English as tea — which, like willow-pattern ware itself from the late nineteenth century, was increasingly coming from the British Empire rather than China.[32] Advertisements in Hawera attest to its popularity into the twentieth century.[33] Interwar interest in *chinoiserie* reached new levels thanks to a series of New Zealand exhibitions of 'Oriental Art' held from the late 1920s to the late 1930s by enthusiastic Sinophiles.[34] Biographer E. H. McCormick (1906–1995) captured the 'cult of eclectic orientalism' encouraged by these exhibitions, which he witnessed while a student at Victoria College (University of New Zealand), Wellington. Along with salvaged Japanese prints 'mounted on strips of fabric and hung over black divans in dimly illuminated studio-bedsitters', McCormick recalled '[r]espectable virgins' ransacking 'the Chinese shops in Wellington's red-light district for rice bowls and fish plates of approved design'.[35]

Other cultural activities in New Zealand also celebrated the willow-pattern design. In 1931, Auckland's League of New Zealand Penwomen put on an 'interesting programme of Chinese music', as the local newspaper reported. This included a 'playlette' — 'The Legend of the Willow Pattern Plate,' — 'produced by Mrs. Culford Bell, with incidental music composed and played by Mrs. Robert

FIGURE 3. *This willow pattern soup plate was unearthed in an archaeological dig on the site of a homestead in New Plymouth burned during the First Taranaki War in 1860. Archaeologists believe that its discovery indicates the popularity of the design among the labouring classes. Source: From the collection of Puke Ariki, New Plymouth, PA2010.029. © Collection of Puke Ariki, New Plymouth. Reproduced by permission of Puke Ariki. Permission to reuse must be obtained from the rightsholder.*

Neil'.[36] In 1936, the *New Zealand Herald* asked its readers to render 'the story of the Willow Pattern Plate in verse form'. Such were the number of submissions, it had to carry selected poems over two separate issues — a few lines from one of which is reproduced at the start of this section.[37] At no less an event than New Zealand's Centennial Exhibition, in 1939, too, schoolchildren performed two plays, one of which was, '"Turtle Dove", a Chinese fantasy'. Here:

The background of the scene was a Chinese garden of the conventional blue willow pattern, the settings being attractively designed and the players, pupils of

Queen Margaret College, were in silken Mandarin robes of Chinese blue with contrasts of a pale shade, also conventional coolie hats, and long pigtails.[38]

A three-dimensional rendering of the willow-pattern plate — in this instance, on stage — was probably not the first to have occurred in New Zealand.

In a further complication of the origins and adaptations of a European idea of a Chinese garden, the *Wellington Independent* of 1869 reported on the construction of a 'tea-garden' by two Chinese. According to it, 'every foot' of land 'seems to have been carefully turned over, and a convenient device has been made in the form of an artificial pond in the centre of the ground'.[39] Sadly, no further record can be found of this garden. Tea gardens — like this one developed by 'our Celestial friends', as the report described the two Chinese — while a popular form of entertainment in colonial society, were largely divorced from attitudes towards Chinese, which by the late 1870s were increasingly hostile.[40]

Willow pattern realized, 1968–2014

In making a three-dimensional representation of the willow-pattern plate, the designer of Hawera's Willow Pattern Garden — Englishman Harry Beveridge — was mining a deep cultural lode in New Zealand by presenting a design which would have been known to almost all. Beveridge's motive in doing so may have been to make his mark on a garden whose previous Superintendent of Parks and Reserves, Donald Ross, had served for thirty-three years.

The site chosen for the Willow Pattern Garden had already undergone several metamorphoses before Beveridge's design. Only eleven years after the 4.3 hectare King Edward Park — a public garden — opened in 1902, enthusiasts established a Fernery Committee in 1913, with a view to raising funds sufficient for its construction.[41] While war delayed this, the committee's energetic and successful fund-raising[42] enabled work to begin on a small fernery in 1917.[43] Located immediately next to the rock garden, King Edward Park's first fernery consisted of native ferns, shrubs, and flowers.[44]

In 1921, construction began on a fernery of 'a more ambitious scale'.[45] Two 'judges of the garden competitions held by the Hawera Town Planning Association' predicted that, on its completion, 'Hawera will possess a fernery with at least a Dominion reputation'.[46] This was by no means a modest aspiration for a town of over 5489 people in 1921,[47] but it belied the importance of gardens and garden making to local identity and pride. Beautifying societies flourished in New Zealand from the late 1890s, and in some places continued to do so for the next 80 to 90 years. They were commonly associated with civic improvement and progressive leagues, and were often the objects of intense rivalry.[48] Indeed, to this day, the province of Taranaki markets itself as 'The Garden of New Zealand', a characterization originating in nineteenth-century descriptions by settlers of its productive soils and temperate climate.

Ferneries, too, were all the rage in twentieth-century New Zealand. In the 1920s, even local halls were decked out to resemble ferneries when they were used for balls.[49] One of the country's first public ferneries opened in Pukekura Park, New Plymouth (figure 4), and rapidly became a 'sight which no visitor should miss'.[50] New Plymouth, a much older town of 19 200 in 1936,[51] was established on Te Atiawa tribal land in 1839 as a planned settlement. Its park developed into one of New Zealand's loveliest, following the creation of an artificial lake in a bush-clad gully, the planting out of the area, and design of paths from the late nineteenth century. By possessing a fernery, Hawera residents, then, could trumpet their own town's eminence as a floral and garden centre able to rival New Plymouth.

Hawera's Mayor officially opened The Parkinson Bequest Fernery in late February 1923. Built with concrete walls and a glass roof, the Fernery comprised six compartments, including ones devoted to ferns and flowers as well as a mixed room containing both ferns and begonias. The local newspaper described the structure:

> A small flight of steps leads down through a rock garden to the main entrance, the floor of the fernery being 5 ft below the level of the ground. The interior presents a real fairyland. Under the glass roof the archways and terraces make a fitting setting for the ferns and choice plants, and a wealth of soft colouring is provided. The fernery measures 57 ft by 36 ft overall and is divided into six compartments, each of which is entered through a mossy reinforced clay archway.[52]

By the next decade, changing fashions dictated that the Fernery become the Begonia House. In 1937, alterations took place to effect this, although the integrity of the original design remained (figure 5). By the 1950s and 1960s, the Begonia House was run down and in need of repair. The allure to vandals of so much glass required its periodic replacement — one young trouble-maker

FIGURE 4. *Pukekura Park, early 1900s. Unknown Photographer, Pukekura Park (c. 1900–1910) and L. Earp & Co., Scenes of and on Mount Egmont (c. 1903–1904). Source: From the collection of Puke Ariki, New Plymouth, PHO2011_0553a. © Collection of Puke Ariki, New Plymouth. Reproduced by permission of Puke Ariki. Permission to reuse must be obtained from the rightsholder.*

had the misfortune to be apprehended by his father, the local policeman.[53] That requirement and the deterioration of the concrete structure meant that by the 1960s it was, in the words of the Park's historian, David Bruce, 'fast becoming an eyesore and needing repairs considered beyond economic reality'.[54]

The Park's new Superintendent, Beveridge, then, used the existing structure of the Begonia House as the basis for his Willow Pattern Garden (figures 6 to 9):

The walled surrounds of the former fernery proved to be ideal for conversion to the traditional willow pattern surrounds, and on designs originated by

Mr Beveridge concrete blocks were cast to top the walls off, a moon gate was cut and willow pattern castings were etched on one wall to give the oriental atmosphere.

Beveridge removed the glass roof, and set to, to transform the Begonia House into a three-dimensional representation of willow-pattern ware.

While site preparation was in progress, at night School, Beveridge 'moulded the oriental figures and then cast in concrete the finished product'. The 'traditional significance' of each figure, added the newspaper, was acknowledged through its placement 'in accordance with the legend'. '[O]riental-styled bridges, the cave of sanctuary and the escape plan' — all surrounded by an ornamental pond representing the sea over which the lovers escaped — completed the scene. 'More atmosphere was created with a stone lantern, which, inlaid with blue glass and a South Island schist stone with a phosphorescent quality, added to the creative scene.' Beveridge also delicately painted the story's three main characters, placing them on the bridge in flight and pursuit respectively. Details from the willow pattern, embossed on the garden's inner walls, were finished in blue. They depict the house the lovers retreated to (northern wall) and their later transmogrification into doves (western wall).

But what were its origins; what, precisely, had inspired Beveridge to make this garden? Wistful memories of willow-pattern ware from dinners taken on his mother's lap? The willow-pattern legend seared into his memory from childhood reading — perhaps, by recalling *Kéramos* by Henry Wadsworth Longfellow (1807–1882)?

The willow pattern, that we knew
In childhood, with its bridge of blue
Leading to unknown thoroughfares;
The solitary man who stares
At the white river flowing through
Its arches, the fantastic trees
And wild perspective of the view;
And intermingled among these
The tiles that in our nurseries
Filled us with wonder and delight,
Or haunted us in dreams at night.[55]

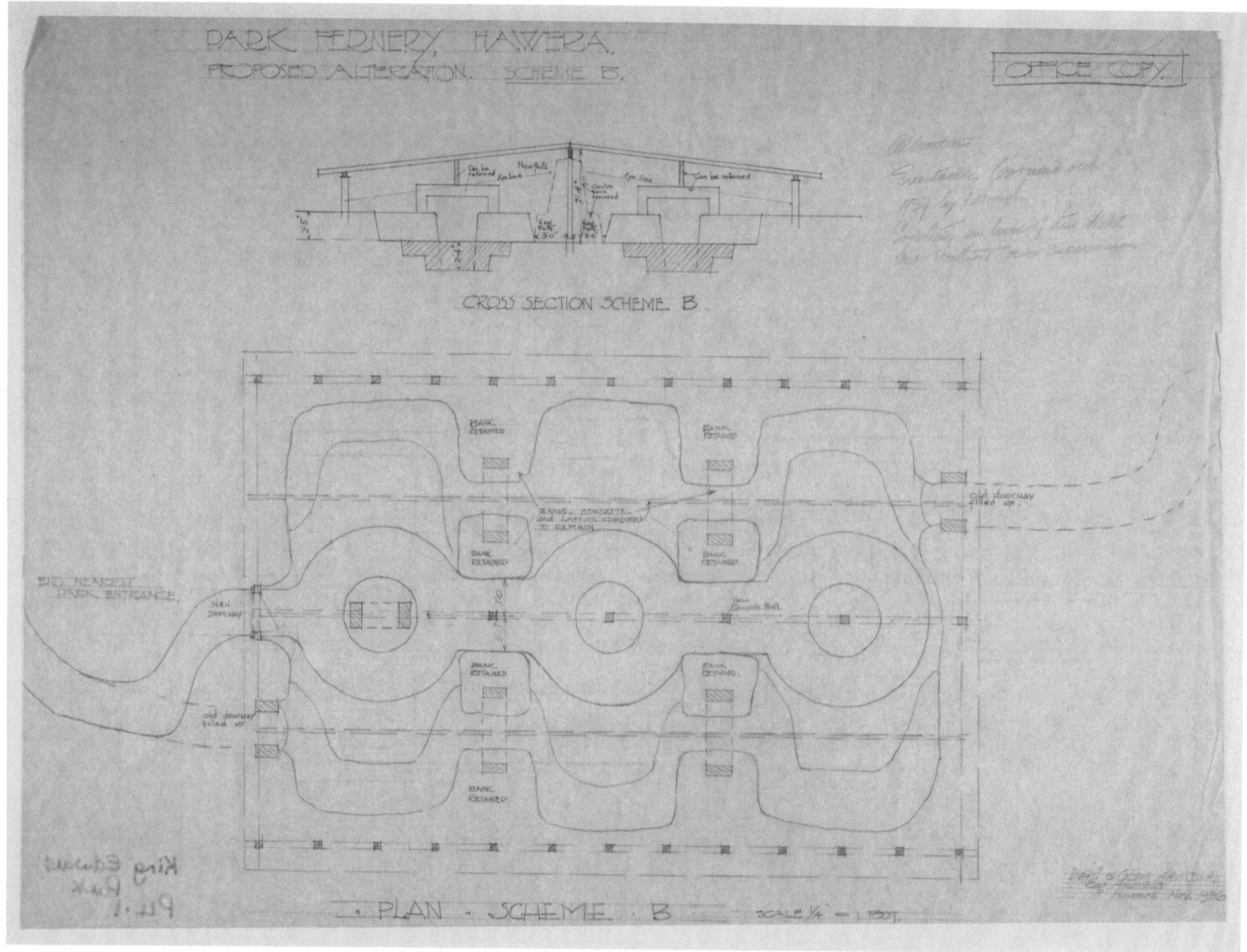

FIGURE 5. *Alterations eventually carried out to King Edward Park Fernery. Rough & Duffill,* King Edward Park Fernery, Proposed Alterations *[Plans] (1936). Source: From the collection of Puke Ariki, New Plymouth, ARC2004–1334. © Collection of Puke Ariki, New Plymouth. Reproduced by permission of Puke Ariki. Permission to reuse must be obtained from the rightsholder.*

FIGURE 6. *Hawera's Willow Pattern Garden, c. 1970s.* © *David Bruce. Reproduced by permission of David Bruce. Permission to reuse must be obtained from the rightsholder.*

FIGURE 7. *Hawera's Willow Pattern Garden through the Moon Gate, 2008. Source: photograph by author.*

Far more prosaically, the concept 'had its foundation', as Beveridge recalled, 'in a floral festival at Oamaru [North Otago, South Island] several years ago, when … [he] was called on to present an interior design for a willow pattern garden display'.[56]

Over the next two decades or more, the Willow Pattern Garden became a showpiece at King Edward Park, featuring regularly in tourism and parks promotions material. Today, older residents recall with pleasure visiting the garden or taking families out on picnics.[57] Its fate, however, followed that of the Fernery and Begonia House. Just as the mandarin's jewellery had proven too great a temptation to Chang, so did the Chinese figures to local vandals and collectors — within a few years they had entirely disappeared. Gradually, the garden fell into ruin, a fall precipitated by the management of the park by a private company and its neglect of the area. By the early 2000s, the garden was in a sorry state.

An oriental garden

But decline has again been arrested, in this case by stimulating the most recent remaking of the Garden in 2014, by Senior Parks Assistant Kane Rangi, who produced a design combining features of both the fernery and the more-recent Willow Pattern Garden. Although this new 'Oriental Garden' retains a Chinese theme, Rangi's design also returns to the layout of the original fernery, whose floor — formerly the Willow Pattern 'Sea' — becomes a sunken path. The new path snakes around plantings that in the Willow Pattern Garden had formed the island and the walkway around the lake (figure 10).

Several motifs from the previous garden remain, to which Rangi has added a red-latticed pergola and Chinese plantings. Most notably, the two doves in relief are now painted in gold set against a red background, a colour-scheme also used elsewhere. Pale blue is kept for the rest of the Garden's inner walls.

FIGURE 8. *Detail of willow pattern story, Willow Pattern Garden Hawera. Source: photograph by author.*

Rangi has added several new features and elements that play on the Willow Pattern Garden. The first is a red lattice-work pergola (figure 11). Nestled in the corner of the eastern- and northern-facing walls, it recalls the blue pergola that previously covered the eastern gate and which now forms part of the walkway winding through the garden. The second new feature introduced by Rangi is the garden's plantings, all of which are Chinese. This includes various flowering cherries, several *Daphne* species, including *Daphne bholua* (the Himalayan flowering shrub), as well as irises and pines, including *Pinus mugo*. The Oriental Garden is also embowered with the Chinese *Ginkgo biloba* and *Davidia involucrata*, the former planted in 2006 as a symbol of the sister city relationship between Harbin (northern China) and South Taranaki, of which Hawera is the administrative centre.

The iconography of the willow-pattern design remains now as a subtle backdrop only, a ghost form tracing the general pattern of a former garden.

Most visitors under a certain age would probably not recognize the spectral outlines of the willow-pattern design, but they would nevertheless acknowledge the general 'Asian' feel it still lends to the garden site.

Conclusion and coda

The creativity latent in the willow-pattern design remains. Recent New Zealand artists have injected new meanings and elements into the willow-pattern design — in doing so, illustrating the same sense of boundless adaptability and creative energy the design inspired in Britain some 200 years ago and, more recently, in Hawera some 50 years ago. One artist to do so is Kerry Ann Lee, a New Zealander of Chinese descent, whose collages use the basic form of the willow-pattern design, but subvert it as a way of challenging 'conventional understandings of identity and belonging' in New Zealand.[58]

As she notes, European occupation of land and identity making have long dominated New Zealand political and cultural life to the extent of excluding the presence of Chinese, who also arrived in the nineteenth century.[59] Lee uses the colour gold, as well as the Chinese bowl and the lantern, as motifs that:

> [C]ondense distances between a migrant heritage from Southern China and the present realities of Chinese New Zealanders. Each of these metaphors embodies aspects of a contemporary Chinese New Zealand identity, while collectively representing both the boon and burden of settlement — a yearning to retain Chinese culture and customs as well as forging a comfortable place in Kiwi society.[60]

For example, her *Blue Willow Landscape* (2007) uses motifs from the willow-pattern design painted over a print of a Charles Blomfield (1848–1926) landscape, *Jackson Peaks, Lake Manapouri* (1912) — a romanticized view of the lake and the Southern Alps painted very much within established European pictorial conventions. Through her collage, Lee 'reclaims' Blomfield's landscape and the willow-pattern design in order to 'announce different cultural understandings of land'.[61]

A modified willow-pattern design also appears in her digital collage, *Lantern* (2007). For Lee, the lantern is a means of drawing attention to the greater public visibility of Chinese in New Zealand. In this instance, she attempts to evoke 'an older Chinese presence that is specifically located in New

FIGURE 10. *Assistant Park's Curator Kane Rangi's sympathetic redesign of 2014 retained several elements of the Willow Pattern Garden. Source: photograph by author.*

Zealand'.[62] Lee's lantern uses the same design and colours as the willow-pattern plate, but depicts recent twentieth-century New Zealand suburban houses rather than the fanciful ones of the original pattern. The willow-pattern lantern appears at the right of the 270 × 700 mm print, which has the Chinese symbol for longevity — *shou* — repeated in white, on a red background. The two doves escape from the lantern onto the mauve background while the phrase, 'Light is good from whatever lamp it shines' appears in white to the left of the lantern.[63] *Lantern* evokes the hitherto invisibility of the Chinese presence in New Zealand, through their making of home and attendant search for stability and good health in that country.

Like Lee, the New Zealand ceramic artist Carole Prentice plays with the willow-pattern design, but instead of a contemporary theme, she has given it a uniquely Pacific one, using porcelain as the medium. Inspired by the delight her children took in reciting a spelling exercise 'that took the form of a cheeky

FIGURE 11. *The new pergola, Willow Pattern Garden, Hawera. Source: photograph by author.*

FIGURE 12. *Recent New Zealand artists have injected new meanings and elements into the willow-pattern design: Carole Prentice's* Maori Willow Pattern *design exemplifies this. Source: Working drawing by Carole Prentice © Studio Ceramics Ltd. Reproduced by permission of Carole Prentice and Phillipa Croft, of Studio Ceramics Ltd, http://studioceramics.co.nz/. Permission to reuse must be obtained from the rightsholder.*

is a Maori *pa* (meeting house), flanked by iconic native New Zealand plants. To the left of the pa is *Tī kōuka* (*Cordyline australis*); behind the pa and to its right, depictions of a mature and youthful New Zealand *Kahikatea* (*Dacrycarpus dacrydioides*); immediately behind the youthful Kahikatea, *Harakeke*/New Zealand flax (*Phormium tenax*). In the middle-foreground, a gnarled *Kowhai* (genus *Sophora*) — drawn in Chinese style — replaces the willow of the classic pattern. A Maori warrior and his lover trade places for Chang and Koong-se. The Pacific translation is complete as the two escape on a *waka* (canoe), construct an early colonial-style house, and finally are transmogrified into a dove and New Zealand *Tūī* (*Prosthemadera novaeseelandiae*) respectively.

These recent manifestations of *chinoiserie* point to the on-going appeal — and eminent adaptability — of a design that in New Zealand has been used for political protest and entertainment, as assertions of identity, and, not least, has been represented in three-dimensions as a garden. The complexity and cultural fusion expressed by the willow-pattern design in New Zealand problematize Pratt's conceptualization of encounter as essentially agonistic and unequal. Instead, they support David Porter's observation on the eighteenth-century introduction of Chinese gardens in Britain — namely, that '[t]he process by which one culture finds meaning in another, rather, entails adaptive strategies that are themselves potentially transformative'.[65]

Acknowledgements

I am indebted to the help of many individuals. Duncan M. Campbell and Richard Bullen generously commented on drafts of this paper, while the two anonymous reviewers made helpful suggestions. South Taranaki District Council Parks Curator David Bruce showed me around the King Edward Park, Hawera, sharing unpublished material and commenting on an early article draft. The editor of the *Taranaki Daily News*, Cliff Hunt, allowed me free and unfettered access to the clippings room and hard-copies of the *Hawera & Normanby Star*. Phillipa Croft, of Studio Ceramics Ltd., and the artist Carole Prentice generously gave me permission to reproduce an image of Carole's fine work, while Carole also answered my several queries about the inspiration behind her work. Fiona Greenhill and Kaye Lally, of Hawera Library, kindly checked for material and introduced me to knowledgeable locals, and the staff of New Plymouth's magnificent Puke Ariki Museum went out of their way to

love song' sung in Maori and English, her '*Maori Willow Pattern* similarly fuses different but familiar visual elements' (figure 12). It reinvents 'in design the traditional Chinese tale of thwarted young lovers who, pursued by outraged relatives, escape to an island in the Pacific where they establish a home and find happiness "underneath the kowhai tree"'.[64] Instead of a Chinese pagoda, there locate material for me. Alan Kynaston drew the wonderful map and diagram accompanying this article and Sarah-Mae Berry provided copy editing.

Disclosure statement

No potential conflict of interest was reported by the author.

Funding

This project and special issue were funded by a University of Waikato Faculty of Arts and Social Sciences Contestable Research Grant.

NOTES

1. Punctuation added. Kerry Ann Lee, *Home Made: Picturing Chinese Settlement in New Zealand* (MA thesis: Massey University of Wellington, 2008), p. 10.

2. David Porter, *The Chinese Taste in Eighteenth-century England* (Cambridge: Cambridge University Press, 2010), pp. 27–28.

3. Mary Louise Pratt, 'Arts of the Contact Zone', *Profession*, xci, 1991, pp. 33–40.

4. Konsin Shah went on to play a prominent role in the diplomatic and political history of Taiwan, his own life providing a fascinating history of China in the second half of the twentieth century, and deserving of a study in its own right. After active service as a pilot fighting the Japanese, he was responsible for flying Chiang Kai-shek and family members from PRC-controlled China to Taiwan. He remained a staunch supporter of the KMT leader. It was his fate, however, to experience the vicissitudes of the ROC's changing diplomatic fortunes, his career peaking just as foreign countries were turning to the PRC. After serving in New Zealand, Shah served as the last ROC Ambassador to the USA, an announcement made when he was in this position and one accepted with shock but characteristic dignity. Later, in Taiwanese politics he served as a staunch supporter of the legacy of Chiang Kai-shek. For further information see: David Dean, *Unofficial Diplomacy: The American Institute in Taiwan: A Memoir* (No place [USA]: Mary Dean Trust, 2014); 'My Cousin Konsin Shah (夏功权)', translated by Sun li, xgls.vicp.net/xgls/小港李氏研究专著/李名宪/06-e-konsin_shah%5B1%5D1.html.

5. For an amusing discussion of this period, see Austin Mitchell, *The Half Gallon Quarter Acre Pavlova Paradise* (Christchurch: Whitcombe and Tombs, 1972).

6. Population from Ron Lambert, 'Taranaki Places — Hāwera', *Te Ara — the Encyclopedia of New Zealand*, http://www.TeAra.govt.nz/en/taranaki-places/page-6.

7. For this pioneering period, see Rollo Arnold, *Settler Kaponga: 1881–1914 — A Frontier Fragment of the Western World* (Wellington: Victoria University Press, 1997).

8. 'Chinese Ambassador Opens Willow Pattern Garden in Hawera'. *The Hawera Star* (21 October 1968), p. 2.

9. David Porter, 'Beyond the Bounds of Truth: Cultural Translation and William Chambers's Chinese Garden', *Mosaic: A Journal for the Interdisciplinary Study of Literature*, xxxvii/2, 2004, pp. 41–58.

10. Anne-Marie Brady, *Making the Foreign Serve China: Managing Foreigners in the People's Republic* (Lanham: Rowman & Littlefield, 2003); Susie Ong, 'Between Great Britain and the USA: New Zealand and the Recognition of China in the 1950s', in Yongjin Zhang (ed.), *New Zealand and Asia: Perceptions, Identity and Engagement* (Auckland: Asia2000 and New Zealand Asia Institute, 1999), pp. 77–98; Chris Elder (ed.), *New Zealand's China Experience: Its Genesis, Triumphs, and Occasional Moments of Less than Complete Success* (Wellington: Victoria University Press, 2012); John McKinnon, 'Breaking the Mould: New Zealand's Relations with China', in Bruce Brown (ed.), *New Zealand in World Affairs, 1972–1990*, 3 vols (Wellington: Victoria University Press, 1999), pp. 226–266.

11. *New Zealand Herald* (29 February 1936), p. 4.

12. David Porter, *Ideographia* (Stanford: Stanford University Press, 2001), p. 135.

13. Porter, *The Chinese Taste*, p. 4.

14. Porter, *The Chinese Taste*, p. 4.

15. Oliver Impey, *Chinoiserie: The Impact of Oriental Styles on Western Art and Decoration* (Oxford: Oxford University Press, 1977), p. 10.

16. Impey, *Chinoiserie*, p. 9.

17. Impey, *Chinoiserie*, p. 14.

18. Robert Finlay, 'Pilgrim Art: The Culture of Porcelain in World History', *Journal of World History*, ix/2, 1998, p. 187.

19. Finlay, 'Pilgrim Art', p. 155.

20. Finlay, 'Pilgrim Art', p. 168.

21. See Hilary Young, 'Manufacturing Outside the Capital: The British Porcelain Factories, Their Sales Networks and Their Artists, 1745–1795', *Journal of Design History*, xii/3, 1999, pp. 257–269. An entertaining discussion of the accidental discovery of porcelain manufacturing in Europe is provided by Janet Gleeson, *The Arcanum: The Extraordinary True Story of European Porcelain* (London: Bantam, 1998).

22. John R. Haddad, 'Imagined Journeys to a Distant Cathay: Constructing China with Ceramics, 1780–1920', *Winterthur Portfolio*, xli/1, 2007, p. 63.

23. Haddad, 'Imagined Journeys', p. 63.

24. *Evening Post* (11 August 1928), p. 20.

25. Carter Harrison, *A Race with the Sun* (New York: Putnam's, 1889), p. 130, cited in Haddad, 'Imagined Journeys', p. 64.

26. Haddad, 'Imagined Journeys', pp. 65–66.

27. Matthew Campbell and Louise Furey, 'Archaeological Investigations at the Westney Farmstead, Mangere', Report to the New Zealand Historic Places Trust, 14 December 2007, p. 99.

28. Viewed in display at Otago Museum, Dunedin, New Zealand. I am indebted to Tony Ballantyne for alerting me to this.

29. Martin Wallace with photography by Mark Forster-King, *Collecting Antique British Blue and White Ware in New Zealand* (Hamilton, New Zealand: Martin Wallace & Mark Forster-King, 2013). See also, Angela Middleton, *Te Puna — A New Zealand Mission Station: Historical Archaeology in New Zealand* (New York: Kluwer Academic/Plenum Publishers, 2008), pp. 202, 206–207.

30. Wallace and Forster-King, *Collecting Antique British Blue and White Ware in New Zealand*.

31. Geometria, *Archaeological Excavation Report on the Street Homestead, Penrod Drive, Bell Block, Taranaki* (Auckland: Geometria Limited, 2008), p. 72. Twelve items of willow-pattern ware, and one of Broseley, were found out of a total of 54 items of crockery. See pp. 71–72. The Spode factory used the pattern name Broseley to describe a design which resembled willow pattern but which composed the scene differently and used a different border.

32. Duncan Campbell, 'What Lies Beneath those Strange rich Surfaces?: Chinoiserie in Thorndon', in Charles Ferrall, Paul Millar and Keren Smith (eds), *East by South: China in the Australasian Imagination* (Wellington: Victoria University Press, 2005), pp. 173–189; Tony Ballantyne, 'India in New Zealand: The Faultlines of Colonial Culture', in Sekhar Bandyopadhyay (ed.), *India in New Zealand: Local Identities, Global Relations* (Dunedin: Otago University Press, 2010), pp. 21–44.

33. *Hawera & Normanby Star* (14 November 1924), p. 1.

34. On which see James Beattie and Lauren Murray, 'Mapping the Social Lives of Objects: Popular and Artistic Responses to the 1937 Exhibition of Chinese Art in New Zealand', *East Asian History*, xxxvii, 2011, pp. 39–58.

35. E. H. McCormick, *The Inland Eye: A Sketch in Visual Autobiography* (Auckland: Auckland Gallery Associates, 1959), p. 26.

36. *Auckland Star* (27 March 1931), p. 10.

37. *New Zealand Herald* (29 February 1936), p. 4; *New Zealand Herald* (7 March 1936), p. 4.

38. *Evening Post* (8 December 1939), p. 11.

39. *Wellington Independent* (17 April 1869), p. 2. I thank Esther Fung for alerting me to this article.

40. James Ng, *Windows on a Chinese Past: How the Cantonese Goldseekers [sic] and Their Heirs Settled in New Zealand* (Dunedin: Otago Heritage Books, 1993), Vol. 1. On Chinese gardens, recent and historic, in New Zealand, see James Beattie, '"The Empire of the Rhododendron": Reorienting New Zealand Garden history', in Tom Brooking and Eric Pawson (eds), *Making a New Land: Environmental Histories of New Zealand* (Dunedin: Otago University Press, 2013), pp. 241–257, 365–367; James Beattie and Duncan Campbell, *Lan Yuan* 蘭園: *A Garden of Distant Longing* (Dunedin: Shanghai Museum Press and Dunedin Chinese Gardens Trust, 2013); James Beattie, 'Making Home, Making Identity: Asian Garden-making in New Zealand, 1850s–1930s', *Studies in the History of Gardens & Designed Landscapes*, xxxi/2, 2011, pp. 139–159; James Beattie, 'Growing Chinese Influences in New Zealand: Chinese Gardens, Identity and Meaning', *New Zealand Journal of Asian Studies*, ix/1, 2007, pp. 38–60.

41. 'King Edward Park reserve was originally a little under 12 ha but had the corners nibbled to provide land for the High School, aquatic centre and motor camp. The latter two and the Hub/Hicks Park sports' complex, which was the former A&P reserve, bringing the total park area now to a touch under 20 ha. King Edward Park (ornamental) is 4.3 ha with another 2.7 ha in the KEP cricket field.' David Bruce, 10 March 2015, email.

42. *Hawera & Normanby Star* (14 April 1913), p. 8.

43. *Hawera & Normanby Star* (9 August 1917), p. 6.

44. *Hawera & Normanby Star* (30 November 1920), p. 4.

45. *Hawera & Normanby Star* (30 June 1921), p. 7.

46. *Hawera & Normanby Star* (19 December 1921), p. 4.

47. Census figures did not list Maori. 'Counties. — Estimated Population, 1st April, 1922', in *The New Zealand Official Year-Book, 1923*, http://www3.stats.govt.nz/New_Zealand_Official_Yearbooks/1923/NZOYB_1923.html.

48. On which, see Thelma Strongman, *City Beautiful: The First 100 Years of the Christchurch Beautifying Association* (Christchurch: Clerestory Press, 1999).

49. *Hawera & Normanby Star* (2 April 1924), p. 8; *Hawera & Normanby Star* (1 September 1921), p. 8.

50. *Auckland Star* (29 February 1936), p. 2.

51. *New Zealand Year-Book, 1936*, http://www3.stats.govt.nz/New_Zealand_Official_Yearbooks/1936/NZOYB_1936.html. By this period, the population figures included Maori.

52. *Hawera & Normanby Star* (28 February 1923), p. 7.

53. David Bruce, 'A Brief History of King Edward Park', unpublished draft. I thank the author very much for making this available to me.

54. Bruce, 'A Brief History', no page.

55. Henry Wadsworth Longfellow, Horace Elisha Scudder (ed.), *The Complete Poetical Works of Henry Wadsworth Longfellow* (Boston and New York: Houghton, Mifflin & Co., 1893), www.bartleby.com/356.

56. *Daily News* [Hawera] (30 August 1968), no page.

57. https://alithena.wordpress.com/author/alithena/page/3/.

58. Lee, 'Home Made', p. 5.

59. On which, see Beattie, '"Hungry Dragons": Expanding the Horizons of Chinese Environmental History — Cantonese Gold-miners in Colonial New Zealand, 1860s–1920s', *Environmental History Review*, 1, 2015, forthcoming.

60. Lee, 'Home Made', p. 21.

61. Lee, 'Home Made', pp. 24–25.

62. Lee, 'Home Made', p. 48.

63. Lee, 'Home Made', p. 48, figure 40.

64. Carole Prentice, 18 February 2015, email.

65. Porter, 'Beyond the Bounds of Truth', p. 56.

Zheng Yuanxun's 'A Personal Record of My Garden of Reflections'

TRANSLATED AND INTRODUCED BY DUNCAN M. CAMPBELL

Making use of the natural scenery is the most vital part of garden design. There are various aspects such as using scenery in the distance, near at hand, above you, and at certain times of the year. But the attraction of natural objects, both the form perceptible to the eye and the essence which touches the heart, must be fully imagined in your mind before you put pen to paper, and only then do you have a possibility of expressing it completely.

> Ji Cheng 計成, *Yuanye* 園冶 [The Craft of Gardens] (1635)[1]

The Bookroom of the Twinned Paulownias, the former garden of the Wang family, restored and expanded by the Zhangs of Guanzhong, is found along Guard of the Left Street … . When I visited Yangzhou during the early years of the reign of the Jiaqing emperor [1796–1820], the owner of the garden, Master Zhang Qinxi, would often host parties there and for a while the garden became the very byword for the pleasures of wine and poetry. Now, more than thirty years later, all the pavilions and terraces seem deserted and bleak, the trees and flowers grow tangled and untended. Is it perhaps the case that the waxing and waning of a garden too is entirely a matter of fate?

> Qian Yong 錢泳, *Luyuan conghua* 履園叢話
> [Conversations from Within Footsteps Garden] (1838)[2]

Translator's introduction

'Ming gardens were social spaces, celebrated in some ways not so much for themselves as for the social gatherings which took place there', argues Craig Clunas.[3] In the spring of the thirteenth year of the reign of the Chongzhen emperor (1640) of the Ming dynasty (1368–1644), some three years after Zheng Yuanxun[4] finished an account of the garden that he had built for himself just south of the city of Yangzhou, entitled 'Yingyuan ziji' 影園自記

[A Personal Record of My Garden of Reflections] and translated below, and some five years after the completion of the garden itself, he hosted there one of late imperial China's most celebrated social gatherings, a party occasioned by the flowering of a single unusual yellow tree peony.[5] In time-honoured manner, the assembled luminaries sought to immortalise the moment (and, by association, the garden, its master, and his guests) by composing poems to the topic 'The Yellow Peony of the Garden of Reflections'.[6] A collection of more than 100 of the poems either written on that day or later solicited from scholars throughout the district was then sent off to be judged by Qian Qianyi 錢謙益 (1582–1664),[7] the pre-eminent if later somewhat problematic literary figure of the day, before being published under the title: *Yingyuan yaohua ji* 影園瑤華集 [The Jasper Flower Collection from the Garden of Reflections].[8]

By Zheng Yuanxun's own account, as we read below, his garden was occasioned by failure. It was designed, he claims, therefore, to provide him with the site wherein he could disengage from society and devote himself in equal measure both to the care of his elderly mother and to his books. To the extent that this claim is not entirely disingenuous — and Zheng seems perhaps somewhat too keen on offering further justifications for his garden (his addiction to the mountains and to the rivers, to bamboo and to trees, his mother's dream, his own destiny) — occasions such as this immortal party appear to have conspired against his desire for anonymous reclusion.[9] Worldly success too was soon to follow the social and cultural prestige thus generated by his garden; by 1643, he had finally achieved success in the Metropolitan Examinations. Ironically, however, it was now to be the circumstances of a dynasty in the final stages of terminal collapse that cast its shadow over the life of the Master of the Garden of Reflections. In 1644 when Yangzhou was surrounded by the army of Gao Jie 高傑 (d. 1645),[10] a former bandit chief now turned Ming loyalist, the populace of the city barred its gates against him.[11] It fell to Zheng Yuanxun to attempt to mediate the crisis. In his 'Hongguang yiyou Yangzhou chengshou ji lue' 弘光乙酉揚州城守紀略 [A Brief Account of the Defence of the City of Yangzhou in the Yiyou year (1645) of the Reign of the Hongguang Emperor], the historian Dai Mingshi 戴名世 (1653–1713)[12] provides the following account of the outcome of his attempt:

> Zheng Yuanxun, an Advanced Scholar of River Capital, a proud and somewhat peevish man, ventured out to parley with the troops. Having entered Gao Jie's encampment, Zheng then proceeded to dine and discourse with him in great merriment, Gao for his part rewarding Zheng for his efforts with gifts of pearls and silk. Upon his return to the city, Zheng appeared even more proud of himself, and addressed the crowd in the following manner: 'General Gao has been summoned here by imperial command. If he was simply obeying orders when he entered Nanjing, how much more so is this the case with Yangzhou?' Aroused to great anger by what they understood to be Zheng's betrayal of the city in order to boost his own reputation, the crowd slaughtered him on the spot and proceeded to eat up his flesh.[13]

Notice of Zheng's first official posting, as Secretary in the Bureau of Operations in the Ministry of War, arrived in Yangzhou three days later.

The bulk of Zheng Yuanxun's account of his garden provides readers with a tour of the various sites of the garden.[14] Like the garden itself, this tour wraps back around itself; we begin our stroll at the outer gate and by the end of the tour we see again the tree growing beyond a wall that is visible from this spot. What lends Zheng's garden an especial level of interest is the involvement in its design of Ji Cheng, the pre-eminent garden designer of the age and author of the single most important traditional Chinese monograph on the craft.[15] In both his 'Foreword' (tici 題詞) to this work, dated 1635 and written in the Garden of Reflections, and, as we see below, in his account of the garden itself, Zheng Yuanxun is explicit about the extent of Ji Cheng's involvement in the design of the Garden of Reflections:

> The most unfortunate thing is if the landowner has the hills and valleys in his heart but cannot express his concept to the workmen, while the workmen can follow instructions but are not creative, and just stick to their plumb-lines and ink-marks. If they thus force the owner to abandon his original concept of hills and valleys to follow their ideas, is that not a great pity? But Ji Wupi [Ji Cheng] has changed all that: he goes by the concept, not by a fixed set of rules, something which most people cannot achieve. And he is even better at directing operations successfully, so that the stubborn becomes flexible and the blocked flows freely: this is really something to be glad about. I am one of Wupi's oldest friends, and I know that he often feels frustrated that a remnant of water and a broken-off piece of mountain give no scope for his accumulated skills; he would dearly love to set out all the ten great mountains of China in one area, and direct a squad of all the mighty labourers of the empire; and to collect together all sorts of exotic, jewel-like flowers and plants, ancient trees and sacred birds to be arranged by him, giving the whole earth a totally new appearance. But alas, there is no landowner with sufficiently grand ideas!

> Does this mean then that Wupi can operate only on a grand scale and not on a small one? No, this is not the case. Different things are suitable both for different gardens and for different people, and no one is Wupi's equal in making appropriate use of what is available.

> When I was building a mansion to the south of the city-wall of Jiangdu, among reedy marshes and banks of willows, the site was only a few yards wide in either direction, but Wupi had only to make some simple arrangements, and it became a magical secluded retreat. I can claim to know a little about garden design and construction myself, but besides Wupi, I feel as clumsy as a cuckoo that cannot even build its own nest.[16]

The tension that runs throughout Zheng's account of his garden and which doubtless dominated the almost decade-long process of its design and construction was that between the physical constrictions of the site (both its overall size and its lack of a mountain) on the one hand, and, on the other, the desire to achieve 'understated elegance of the simple and the rustic' (puye zhi zhi 朴野之致) whilst avoiding any anxiety about the exhaustibility of the delights it contained as one moved through it.[17] To this end, the design of the garden embodies many of the conceits of concealment and disclosure, surprise and anticipation that feature prominently in Ji Cheng's monograph.[18] The extent to which the Garden of Reflections sought also to borrow from beyond its own walls views of sites of scenic beauty and historical importance was also, in Ji Cheng's terms, a true measure of the craft.[19] Significantly, the aesthetic sensibilities of both the master of the garden and those of his friend, the master of the craft, derived quite explicitly from their study of the traditions of landscape painting.

Sadly, the Garden of Reflections appears not to have survived much beyond the dynastic transition that had so tragically entangled its master, although the family tradition of garden building seems to have been maintained by succeeding generations.[20] Writing sometime around 1691, for instance, Fang Xiangying 方象瑛 notes the disappearance of Zheng Yuanxun's garden in his account of the restoration of a garden then in the possession of Zheng's grandnephew Zheng Maojia 鄭懋嘉:

> The prosperous and picturesque suburbs of River Capital once boasted so many gardens, and yet ever since the time of Emperor Yang of the Sui such gardens as 'Firefly Park', 'Labyrinth Tower' and 'The Songs and Pipes of Bamboo West' have all long disappeared without trace. This fate is true even of Master Zheng Yuanxun's 'Garden of Reflections', which was once the splendour of the age. In those days, whenever the yellow tree peony burst into flower, all the most famous literary men from throughout the empire would gather in this garden to eat and drink and to compose poetry. On one such occasion, their poems were gathered up into a bundle that was then sealed up and sent off to Master Qian Qianyi of Yushan for his evaluation. The poems of Li Suiqiu of Nanhai were adjudged the best and he was rewarded with a golden goblet — how very splendid it all was! Now, in the twinkle of an eye, fifty years have gone by; the garden changed hands, its pavilions and gazebos fell into ruin and nobody any longer even knows where the garden had once been. This 'Garden of Rest' once owned by Zheng Xiaru 鄭俠如 of the Ministry of Works alone has survived intact the ravages of war and calamity, proving once again that the vicissitudes of a garden are determined always by the fate of its owner.[21]

Toward the end of an important article addressing aspects of the historiography of the garden in China, Stanislaus Fung discusses what he considers to be a significant 'blackout'.[22] For all the traditional primacy of verbal rather than pictorial representations of the garden in China, a term equating with the concept of garden history is all but absent in the traditional sources. The one instance that Fung identifies is a preface written by Chen Jiru 陳繼儒[23] to the work of a friend entitled *Yuan shi* 園史 [A History of the Garden], found in Wei Yong's 衛泳 late-Ming anthology *Bingxue xie* 冰雪攜 [Portable Ice and Snow]. The beginning of his preface, in Fung's translation, reads:

> I once said that there were four difficulties with gardens: it is difficult to have fine mountains and waters; it is difficult to have old trees; it is difficult to plan; and it is difficult to assign names. Then there are three easy things: the powerful can easily seize the garden; in time, it can easily become unkempt; and with an

uncultivated owner, it easily becomes vulgar. Nowadays, there are many famous gardens in Jiangnan. I often pass by them and rest my eyes on them. However, when I next visit them, they might still have bright flowers and shaded ferns, but the owner would not have the leisure to be there, or even if he could be there, he would fling his arm around and depart like a courier; or he would diminish the plans of his forebears, altering them abruptly each summer so that the garden would not be completely renovated when his bones would already have decayed; or in the twinkling of an eye, he would sell it to another family, and then if a huge plaque does not label the entrance, a strong lock would bolt the door shut; or trees would be cut down to make mortars, and rocks would be pulled down to make plinths. The fallen beams and ruined walls would be like a house abandoned during a drought. Even if the eaves, rafters and shingles are maintained well, and the pines and chrysanthemums are the same as before, the owner of the garden could be an old wine-drinking and meat-eating reprobate; and every fern and every tree, every word and every sentence would cause the viewers to belch and feel like vomiting. They would stop their noses, cover their faces, and could not remain there for another moment. Having it would be a cause for regret; how could this compare to the pleasure of being rid of it.[24]

In Chen Jiru's terms and with the help of both his friend Ji Cheng and Chen's own contribution of the inscription for a plaque, Zheng Yuanxun can be said to have successfully negotiated these 'four difficulties' (*si nan* 四難), whilst his premature death meant that he did not fall prone to Chen's 'three easy things' (*san yi* 三易) either. On the basis of his reading of this passage in particular, Fung goes on to argue that in Chinese terms, 'garden history' constituted 'a poetic record of experiences in a particular garden' (p. 215) and that 'the focus of narration in traditional writings does not shift from particularity toward generalization and abstraction, but broadens to show how gardens were part of the transformation of dynastic fortunes' (p. 216). It is the very ephemerality of gardens (and of their ownership, one might add), Fung concludes, that 'offers the springboard for historical reflections that relate gardens to a broad picture of historical change'. The essential melancholy of his reflection seems to be captured by the commentary attached to Chen Jiru's preface by Wei Yong, the compiler of the anthology in which it was published, writing under the sobriquet 'The Lazy Immortal' (*lanxian* 懶仙):

> This great earth of ours is but a solitary 'grass hut'[25] which, down through the ages, has never had a constant master. I for my part desire no more than to don my straw sandals and take up my bamboo staff to go free and easy wandering all over the place, and to find total contentment in whatever it is that destiny may

bring me. As for that foot of land in front of my gate, that handful of mountain and that ladle full of water, how could I ever possibly manage to possess it for ever? Thinking thus brings tears to my eyes!

It is, however, a melancholy somewhat ameliorated, surely, by the immortality offered both the gardens themselves and their masters by the preservation of the written word which they had served to generate.[26] A couplet from a poem celebrating a tour of the Garden of Rest by Quan Zuwang 全祖望 (1705–55),[27] the Qing dynasty (1644–1911) historian and frequent visitor to Yangzhou, reads:

Those famous flowers to the Heavens have ascended,
But the tall trees survive yet south of the highroad.[28]

To these lines, Quan adds the annotation: 'This is to say that one treasures still the memory of Zheng Yuanxun's Garden of Reflections'.

A Personal Record of My Garden of Reflections

An addiction to the mountains and to the rivers, to bamboo and to trees is something with which one is born; it cannot be affected. Born, as I was, in River North and thus deprived of the sight of even handfuls of stone,[29] throughout my youth it was only in paintings that I came to see tall mountains with lofty peaks, but my love for them proved quite irresistible as I explored them in my mind. Over time I taught myself to paint, my paintings, however, according with no single tradition of the art. Whenever I went beyond the suburbs, the sight of the delicate beauty of the forests and the rivers would so detain me that I could not bear to return home, and thus it was that during my studies I usually took up residence in deserted temples.

It was only when I was in my seventeenth year that I finally crossed south of the river and visited the various splendours of Jinling. Within the decade, I had visited more than half the splendours of the Wu region, a circumstance that both lent consolation to my innermost desires and served to convince me that there is nothing more fitting in life than such travel. Upon my return, I playfully tried to capture with ink and brush the forms of those splendours I had encountered and when, in the winter of the *Renshen* year [1632], Master Dong Qichang[30] happened to pass through Yangzhou, I took my paintings to him to solicit his criticisms. He, for his part, was kind enough to praise my

skills, saying that as I had managed to capture the essence of the mountains and the rivers, my paintings should not be judged simply in terms of the skill of my brushwork.

Prevailing upon this excess of goodwill, I made bold enough to continue: 'I am now over thirty years old and my life so far has been characterised by a distinct lack of good fortune. My scholarship too proves patchy in the extreme. Recently, however, I happened upon a disused vegetable garden south of the town which I have since acquired and where I intend soon to have constructed a simple cottage of several bays in size, my plan being to live there for the rest of my life, caring for my mother and continuing with my studies. Every now and then, when I find myself free from my various other tasks, I will make copies of the paintings of famous sites by the Masters of old and in this process I hope to be able to make recumbent tours of those places. What do you think of this plan?'

The Master replied: 'This all sounds splendid indeed. But does this site you speak of contain a mountain?'

'No, it doesn't', I admitted, 'and yet to both front and back it is bounded by a river, and across the water Shu Ridge undulates like a writhing serpent with all the force of a mountain. On all four sides grow ten thousand willow trees and more than a thousand *qing* of lotus plants. There too grow bulrushes and the river flows clear and the fish are plentiful, with fishing skiffs coming and going all day long. As spring gives way to summer, people come here to listen to the orioles. Connected as is this site to the very tail of the Sui Embankment and along a somewhat circuitous path, however, infrequent are the visitors who pass by here and thus does it acquire its air of tranquillity. Climbing up to its highest point in order to gaze about one, both Labyrinth Tower[31] and Level with the Mountains Hall[32] appear as if at one's shoulder, whilst each and every one of the green hills of River South loom clearly into sight. The site stands bathed in the reflections cast by the willows, by the river, and by the mountain, and if it has little else to recommend it, it is nonetheless the choicest site of my home district'.

To this, the Master responded: 'It is a site, then, that should afford you both an excess of pleasure and a modicum of consolation', before proceeding to inscribe the words 'Garden of Reflections' to present to me.

My return home in the *Jiaxu* year [1634] coincided with the death of my wife, in addition to which I developed a painful eye infection that left me almost blind. Finding that I could now neither read nor drink, my sense of melancholy grew apace and I all but lost interest in life altogether. My mother became extremely anxious at this turn of events and was forever counselling me to force myself to seek some form of amusement in order to take my mind off my troubles. My brothers too exhorted me to begin the construction of my garden here, and so it was that, having acquired this plot of land some seven or eight years previously and having spent these seven or eight years assembling the materials necessary for its construction, with everything now in readiness and with the conception of the garden fully formed in my mind, my garden began to take rough shape within the space of eight months.

The outer gate of the garden faces east and overlooks the river, with south city standing on the opposite bank and both banks lined with peach and willow trees, their floating reflections stretching away to both north and south. During the months of spring, the people who come here by boat call the place 'Little Peach Blossom Spring'.[33] Once through the gate, a mountain path describes a turn or two within the dense shade cast by the hanging branches of tall pine and fir, and here and there grow flowering prunus, apricot, pear and chestnut trees. Where the mountain ends, to the left, there stands a trellis of rose-leaf raspberry, beyond which stands a clump of bulrushes, this being where the fishermen gather with their nets. A small stream flows along the right hand side, and beyond the stream grow a hundred or so sparsely planted bamboo, shielded by a squat hedge made out of roughly hewn old branches. A surrounding wall has been inlayed with an assortment of unevenly shaped pebbles, these pebbles all having been chosen for their mottled tiger-skin colours, hence the common appellation for this type of wall: 'Tiger-skin Wall'. Two small gates have been formed here out of gnarled tree trunks that resemble coiled dragons in shape. On beyond the gates ten or more tall paulownia trees grow, their branches now woven together in an arch over the path, their backs to the sun as they twist and turn. When people happen to walk along this way, both their gowns and their faces acquire a greenish hue, such is the quality of the shade cast by the trees. Further on again and through the gate one comes upon the plaque inscribed with the words 'Garden of Reflections', this structure being my library. Why have I called this place a garden? In ancient times dependent states were called 'Reflections' and as I am here surrounded, to left and to right, by gardens, so I may be permitted to take this name from this dependency.

Turning into a narrow path, I see branches of flowering plum poke up from behind a wall but I have no idea what it is that lies behind this wall. Across

Willow Embankment, upon 'the clumps and stumps', an aged Chinese trumpet creeper coils its way upwards, its flowers drooping downwards. Where the willows end, one crosses a small stone bridge, crafted from a disorderly pile of rocks, with a tiger lying before it and a recalcitrant stone stretching the entire way across. A turn takes one into the thatched cottage, the plaque of which, inscribed by the Minister of State Zheng Yuanyue 鄭元嶽 of my family, bears the words: 'Cottage of the Jade Hook'; this district once boasted a Jade Hook Grotto Heaven and perhaps this was its original site? The cottage is sited beside the river and is surrounded on all four sides by ponds, all of which are entirely given over to lotus. In design, the cottage is spacious and light, drawing to itself all the ambient blue-green hues and its door and window lintels are made to unusual patterns. Backing on to the cottage is a pond and beyond this pond stretches an embankment upon which grow tall willows, and beyond the willows again is the long river along the opposite banks of which, again, grow tall willows; the gardens once owned by the Yan 閻, the Feng 馮 and the Yuan 員 clans can all be seen as a glance. Although all these gardens now lie in ruins, their trees and bamboo flourish still and it is as if they now form part of my own garden. To the river's south is the transport ford, controlled by the river police. To the north, the river takes one directly to the Old Han Canal, to the Sui Embankment, to Level with the Mountains Hall, to Labyrinth Tower, to Flowering Plum Ridge and to Dogwood Bay, and thus giving the expression 'ten thousand willow trees', the entire scene, from my garden all the way here, appearing like an endless bolt of unravelled embroidery. By nature, the oriole is drawn to the willow and the more the willows the merrier the orioles become; here their song is never ending and many are the listeners drawn to this place. A small belvedere was built overlooking the river, called 'Half Floating' as it was built jutting out over the water, and this has been designated as the place at which to listen to the orioles or from which one can cast off in a little skiff to pay them a visit. These skiffs are the size of the petals of a lotus plant and are called 'Floating Hermitages'; large enough to accommodate a single couch, a small side table, and a tea brazier, in them one can ply one's way to all the local splendours, to Han Canal, to the Sui Embankment, to Level with the Mountains Hall and to Labyrinth Tower, whenever the whim arises.

Beneath the cottage once grew two ancient 'Western Palace' crab-apples, both two *zhang* in height and ten *wei* of girth. I have no idea when they were planted here but they were known to be the only specimens to be found growing throughout the River North region. Only one of these trees now remains, and the sight of it induces in me a sense of intense melancholy. Around the pond, steps of yellow rock have been placed at irregular heights, some of which look like terraces, others as if they are born of the water itself.[34] Ten or so people can stand on the largest of these rocks at any one time, four or five on the smallest, and everybody calls them: 'Little Thousand Men Thrones'. Those rocks with their bases submerged in the water are surrounded by lotus flowers; those on land by flowering plum, magnolia, 'Hanging Floss' crab-apple and yellow and white peach. Orchids of various kinds, Beauty of Yu poppy, ornamental ginger and Sweet William and so on have been planted in the cracks in the rocks. A twisted plank bridge with red handrails takes one across the pond, threading its way through the weeping willows. At the midpoint of this bridge there is a Conning Tower, the bridge itself leading neither to Half Belvedere, Small Pavilion nor to River Belvedere. At the other end of the bridge a stone has been engraved with the words: 'Light Mist and Fine Rain', also in the calligraphy of the Minister of State, but in this instance in a style that closely resembles that of Su Shi[35] of the Song dynasty.

Once through the gate, one finds that a winding gallery forks off to both left and right, the former direction leading to my study. This comprises a three-bay chamber and a reception room, also three bays in size and which, although west facing, protected as it is by paulownia and willow, affords shelter from the summer sun and seems to catch the faintest zephyr. The chamber itself is divided into two sections, one of which faces south, and the door of which is hidden, this being where I escape from my visitors. The windowsill is a full *chi* above the ground so that the chamber remains dry and free of damp. Outside this window is a rectangular porch upon which have been placed a number of large rocks and three or four plantains, along with a single teak tree that came from the western regions, as well as innumerable begonia shrubs, the ground here having been covered in goose egg pebbles. The latticework of the window inside the chamber that opens up to the outside is in the gardenia flower pattern;[36] the window itself is screened off by densely planted bamboo. Even when people happen to catch sight of the window, they can never manage to find the door.

The chamber on the left is east-facing and, when viewed from within the book storage room, the width of the belvedere seems to balance perfectly that of the proportions of the chamber. From here, one can gaze into the far distance upon all the various peaks of River South, taking in the divers hues of the trees both near and far. When bandits threatened the neighbourhood and Salt Commissioner Deng ascended the city walls to inspect the defences, he declared my belvedere tall enough for the bandits to make use of as a lookout if they were to occupy it. Upon hearing this, I had it dismantled in the space of a single evening and later replaced it with a smaller one-bay belvedere which everyone thought even more refined than the original structure. In the forecourt I selected rocks that best embodied the qualities of foraminate structure, leanness and fineness[37] and had them placed at differing heights here and there, again to a design that avoided the fashion of the day but which embodies a painterly quality.

At the corners of the chamber, two steep ridges have been formed, planted mainly in cassia trees and the branches of which over time have intertwined with each other; with its valley stream and precipitous cliffs, the scene resembles that described in Little Mountain's 'Summons for a Recluse'.[38] Tree peony, 'Western Palace' and 'Hanging Floss' crab-apple, magnolia, yellow and white and 'Big Red' pearl camellia, 'Fragrance of the Mouth of the Musical Stone' winter sweet, 'Thousand-leaf' pomegranate, blue-white Chinese trumpet and sweet smelling citron grow beneath the ridges, providing my garden with year-long colour. A single large rock has been placed here to screen the plants, and beneath this grows a solitary old scholar tree, now gnarled and twisted with age; I pat its trunk affectionately, this hundred-year-old tree, and call it my 'Little Friend'. Turning in around the corner of the rock and opening up a small one-leaf door, one comes across a tiny pavilion looking out over the river. Here, the wild rice and the bulrushes form a canopy. My friend and fellow society member Jiang Chengzong 姜承宗 named this pavilion 'Amidst the Wild Rice and Bulrushes', whilst the plaque inscribed sometime earlier by Master Ni Yuanlu[39] with the words: 'Pavilion of the Blue-green Waves' hangs here too. As the autumn grows old, the bulrushes are like snow and the wild geese and ducks make their homes here, departing at dawn only to return at night. Although they keep me company as I study my books I dare not feed or water them. Lying in the pavilion during the dog-days of summer I can enjoy whatever breeze there is to be had, and when the moon rises from the tips of the willow branches, it seems as if newly bathed in an ice pot. As dusk begins to fall, I gaze at the rays of the setting sun as they strike Shu Ridge, the red slowly sinking into the greenness and turning it into an even finer shade of emerald, catching the passers-by in their glare and serving to confuse the homing crows. Although the tiny belvedere is built within the chamber, it cannot be entered from inside the chamber itself and to do so one must take a circuitous route outside the chamber where there is another gallery, to the right of the gate. This gallery winds around twice, and in the gaps are planted mottled bamboo, or plantains, or elm trees, to provide it with shade. Whenever, sitting here within the inner chamber, I feel the desire to mount the belvedere but find myself too lazy to do so, I promptly change my direction and turn inwards. From the gallery inside the door of 'Light Mist and Fine Rain' one enters to the right a covered walkway, in shape like a pavilion, this being the Conning Tower on the bridge. It is also called a 'pavilion' with the proposed name 'Drenched Eyebrow's Prominence',[40] because it overlooks the water as an eyebrow overlooks the eye and because the projecting eaves have been used to form a belvedere. The window has a two-leaf shutter that can be opened or shut whenever required. Two pathways lead away from the back of the pavilion, one of which leads to a hexagonal doorway. Within this door there is a chamber and a reception hall, both of three bays in size, named 'The Studio of the Single Character' and wherein hangs a plaque presented to me by my former preceptor Xu Shuoan 徐碩菴, this being where I teach my sons to read. The reception hall is particularly large and airy and is bordered by a purple balustrade that is exquisite but not at all gaudy. Below the steps grow a single ancient pine and a solitary pomegranate tree. The terrace is shaped like a half sword ring and all around I have planted tree peonies and white peonies. Beyond the low wall one can see a rock face where two pines stretch straight upward, seemingly reaching half the way to heaven. There is another large doorway directly opposite the hexagonal one, beyond which there is also a winding gallery. Here, the dwarf bamboos hug the vermilion balustrade and every now and then the gallery widens or narrows suddenly so as to induce a sense of unpredictability. A single tiny doorway has been retained, through which one can see a cassia tree that looks as if it is the one found growing on the moon, this route providing an alternative way out of my garden. Half Belvedere is sited behind 'Drenched Eyebrow's Prominence', to the left of the path, where, through a wide gallery, one ascends the steps. Hanging here is the plaque once given to me by Master Chen Jiru and inscribed with the words: 'Belvedere of the Love of Solitude'[41] from the line of a poem by the Tang poet Li Bo that goes: 'Overwhelmingly do I love the solitude of seclusion'.[42] The belvedere is surrounded on three sides by water; on the remaining side there is a rock face that seems to surge upwards with a never-ending vigour and on the summit of which are planted two 'toothpick' pines, these being the trees that can be seen when standing in front of 'The Studio of the Single Character' and one of which

looks even more powerful when bent under the weight of a recent downfall of snow. Beneath the rock face runs a stony creek that allows the water of the pond to flow in here with a melodious burble. The creek is edged by large rocks that jut up sharply, in the cracks of which have been planted 'Five coloured' flowering prunus. Having encircled three sides of the belvedere, the creek then disappears into the river, but just before it does so, a single rock rises all alone in the midst of the water and on this too a flowering prunus has been planted, this being the tree that is visible beyond the wall as one first enters my garden.

The back window of the pavilion faces the cottage and people in the cottage and people here can gaze at each other, call out to one another and even converse, without ever discovering the path that leads from one spot to the other. Although, in total, my garden is no wider than several *mu*, visitors to it never fall prey to any sense of anxiety about the exhaustibility of its pleasures. Its mountain paths, both upper and lower, do not criss-cross each other and are thus level and easily walked along, following as they do the natural undulations of the land and giving no visible sign of the effort of man. And yet each and every flower and bamboo and rock seems as if planted in its most natural place; each has been examined and deliberated upon time after time, to be discarded when it did not prove appropriate, regardless of how beautiful it may well have been when examined on its own.

A patch of surplus land has also been retained, a short stroll distant from my garden, and this is put to use as a nursery for the trees and flowers that need to be replaced. There is also a lotus pond several *mu* in size, with a thatched pavilion built upon an outcrop surrounded by water where I can sit and supervise the nurserymen at their work. When the flowers burst into bloom, I ascend the steps of the stone bridge within my garden, or go to Half Belvedere, and from these places I can look out over them. The four or five fishermen who happen to live here have no idea how very fortunate they are. The poet Wang Chun[43] designated this spot as a place for releasing life and named it: 'Tower of the Precious Stamen'; thereafter, the chant of Buddhist sutras would occasionally be heard here. When Wang Chun died, it was here also that his sacrificial rites were conducted and a fellow society member Yan Sheqing 閻舍卿 has protected the site and has maintained it as a place for releasing life. Wang Chun had been my friend in life; he remains my neighbour still and has now become my friend in death.

This project of mine began to take rough shape after eight months; within a year's time the work on it had been completed and it served to overturn completely established patterns of garden design, particularly in its approximation to the understated elegance of the simple and the rustic. Furthermore, it benefited from the intuitive grasp of what I had in mind on the part of my friend Ji Cheng from Wujiang who developed all my ideas to their logical conclusions and whose management of the stonemasons proved so faultless that I had no basis whatsoever to complain about divergence between my plan and the garden as it now is.[44] Sometime previously, my elderly mother had a dream in which she found herself at a site where a garden was being constructed. 'And to whom does the garden belong?', she had inquired, to which the reply had been: 'Your second son'. At the time of her dream, I was still a youth but once I had assembled my workers my mother happened to come to my garden in order to jolly them along in their various tasks. All of a sudden she realised that my garden resembled that of her earlier dream. Once she had told me all about this, I for my part realised that my choice of this particular site had not been at all accidental and that it was unavoidable that I should have embarked upon this present project. And how is one to know that Master Dong Qichang's naming of the garden with the word 'Reflections' was not too a revelation derived from the illusion of a dream and that in my present folly I am not seeking to realise the dream of some earlier man?[45]

It is a truth that men of this present age of ours are wont to compete over that which is real but cast aside that which is illusory, and if one were now to compare this garden with fields and with mansions, then it is the garden that constitutes the illusion. Further, if one were to compare watering the trees and the flowers of this garden with establishing merit and making a reputation for oneself, then surely it is watering one's garden that constitutes the illusion. Before a man takes pleasure in constructing a garden, he should first have acquired fields and mansions and established both his merit and his reputation; has there ever before been a man, such as I, who, without a foot of land to his name, with his mansions not as yet built and with neither merit nor reputation to speak of, first proceeds to build a garden? Were there to have been such a man, he would obviously have brought ruination down upon his own head and dissipated his will for greater things. And yet, to have a mother and not to find the time to care for her, to own books but not to find the time to read them, to be surrounded by the objects that please the senses and satisfy one's nature but never to find the time to enjoy them, is this a circumstance caused by the encumbrance of having a garden that requires watering or rather the ownership of fields and mansions and by the need to establish both one's merit and one's reputation? This is a question to which I dare not offer an answer, although I do believe that each of us must follow the respective dictates of Heaven. The conception of this garden was revealed in a dream, its realisation was the result of my own innate nature. Even if it is the case that I can be accused of preferring to dwell in the illusory rather than the real, then of what possible concern is this to others?

A personal record made by Zheng Yuanxun of Hanjiang[46] during the 2nd Month of the Dingchou year of the reign of the Chongzhen Emperor [1637].

NOTES

1. Alison Hardie, trans., *The Craft of Gardens* (New Haven & London: Yale University Press, 1988), p. 121.
2. Beijing: Zhonghua shuju, Vol. 2, pp. 531–532.
3. 'Ideal and Reality in the Ming Garden', in L. Tjon Sie Fat and E. de Jong, eds, *The Authentic Garden: A Symposium on Gardens* (Leiden: Clusius Foundation, 1991), p. 203.
4. Zheng Yuanxun 鄭元勳 (*zi* Chaozong 超宗, *hao* 惠 東; 1603–44), originally from She County in Anhui Province but a long-term resident of River Capital (Jiangdu 江都), present-day Yangzhou. For a biography, see Hang Shijun 杭世駿 (1696–1773), 'Ming zhifangsi zhushi Zheng Yuanxun zhuan' 明 職方司主事鄭元勳傳 [Biography of Zhang Yuanxun, Secretary in the Bureau of Operations of the Ministry of War of the Ming Dynasty], *Daogutang wenji* 道古堂文集 [The Talking of the Past Hall Collection] (1776), 29: 1a–7b; I am indebted to my colleague Michael Radich for supplying me with a photocopy of this source. For a general treatment of the social and economic context of the Zheng family's wealth, and their manner of spending it, see Ping-ti Ho, 'The Salt Merchants of Yang-chou: A Study of Commercial Capitalism in Eighteenth-Century China', *Harvard Journal of Asiatic Studies*, 17/1–2, 1954, pp. 130–168;

and Antonia Finnane, *Speaking of Yangzhou: A Chinese City, 1550–1850* (Cambridge, MA: Harvard University Press, 2004), esp. pp. 63–68, 80–84. Finnane comments: 'There were gardens in Yangzhou before Zheng's time, but his Garden of Reflections (Yingyuan) marks the proximate beginning of salt merchant gardens in the city' (pp. 64–65). Zheng Yuanxun's 'Yingyuan ziji' 影園自記 [A Personal Record of My Garden of Reflections] was first included in his *Yingyuan yaohua ji* 影園瑤華集 [The Jasper Flower Collection from the Garden of Reflections] (1640) and subsequently much republished in local gazetteers; this present translation is based on the text as reprinted in Chen Zhi 陳植 and Zhang Gongchi 張公弛, eds, *Zhongguo lidai mingyuan ji xuanzhu* 中國曆代名園記選注 [Famous Chinese Garden Accounts Through the Ages: Selected and Annotated] (Hefei: Anhui kexue jishu chubanshe, 1983), pp. 220–227. An earlier version of this present translation was published by the Asian Studies Institute of Victoria University of Wellington as part of the Institute's Translation Papers. I am most grateful to Alan Berkowitz, James Beattie, and Malcolm McKinnon for their comments on earlier versions of this paper.

5. A variety of the *mudan* 牡丹 (*Paeonia suffruticosa*), traditionally regarded as the 'King of the Flowers' (*huawang* 花王). For a traditional discussion of the cultivation of this flower, and a listing of its important varieties, see Chen Haozi 陳淏子, *Huajing* 花鏡 [A Mirror of Flowers] (1688; Beijing: Nongye chubanshe, 1962), pp. 94–100. From the Song dynasty onwards, Luoyang became the most highly regarded centre of tree peony cultivation and the Song historian Ouyang Xiu's 歐陽修 (1007–72) *Luoyang mudan ji* 洛陽牡丹記 [Record of the Tree Peonies of Luoyang], completed in 1034, was the single most celebrated monograph on their cultivation. For a discussion of the tree peony generally and of this work in particular, see Joseph Needham, ed., *Science and Civilisation in China: Vol. 6: Biology and Biological Technology: Part 1: Botany* (Cambridge University Press, 1986), pp. 394–409. 'So things which possess extreme beauty or extreme ugliness are the results of an imbalance in the vital *qi* (皆得於氣之偏也)', Ouyang Xiu argues, in Needham's translation (Romanisation altered), 'The beauty of a flower, the grotesque ugliness of the twisted bulge of a gnarled tree, are indeed different, yet they are equal in their defectiveness, for each comes from an unbalanced quality in the vital *qi* … . Extraordinary things which cause harm to mankind we call calamities (*zai* 災), extraordinary but harmless things which only cause wonder and amazement we call marvels or monstrosities (*yao* 妖). The saying has it: "Heaven contravening the seasons means calamity, Earth conflicting with normal things means a marvel". The tree-peony is truly a bewitching marvel among the plants, and one of the wonders of the ten thousand things. Unlike the twisted bulge of a gnarled tree (its *qi* is unbalanced) only on the side of beauty — therefore it finds favour and blessing among men' (p. 405).

6. The guest list included: Chen Danzhong 陳丹衷, Jiang Chengzong 姜承宗, Mao Xiang 冒襄, Li Zhichun 李之椿, Gu Ermai 顧爾邁, Liang Yusi 梁于涘 Wang Guanglu 王光魯, Li Suiqiu 黎遂球, Cheng Sui 程邃, Wan Shihua 萬時華, and Mao Yuanyi 茅元儀, for which see Zhang Huijian 張慧劍, ed., *Ming Qing Jiangsu wenren nianbiao* 明清江蘇文人年表 [A Chronology of the Activities of the Literati of Jiangsu Province During the Ming and Qing Dynasties] (Shanghai: Shanghai guji chubanshe, 1986), p. 550. In her *A Bushel of Pearls: Painting for Sale in Eighteenth-Century Yangchow* (Stanford, CA: Stanford University Press, 2001), pp. 27–41, Ginger Cheng-chi Hsu discusses the importance of such gatherings with particular reference to Yangzhou during the early years of the Qing dynasty. Of particular relevance to our present context is her translation of a passage from Li Dou's 李斗 (fl. 1795) *Aitang qulu* 艾塘曲錄 [Aitang's (Li Dou) Catalogue of Song Titles]: 'Among literary gatherings of Yangzhou, those taking place in Xiaolinglong shanguan [小玲瓏山

館 (Little Translucent Mountain Lodge)] of the Ma [馬] family, Xiao Garden [篠園 (Garden of the Dwarf Bamboo)] of the Cheng [程] family, and Xiu Garden [休園 (Garden of Rest)] of the Zheng [鄭] family are most active and elaborate. At regular meetings, desks are set up for each guest. On each desk, there are two brushes, inkstones, a water jar, four pieces of writing paper, a note for rhymes, teapot and bowl, fruit and dessert containers, all meticulously arranged. The poems are sent to the carver for publication as soon as they are finished. There is a three-day grace period allowing for participants to revise or edit, but as soon as the poems are in print, they spread to all corners of the city' (p. 29; Romanisation altered; Chinese characters and translation of garden names added). This passage occurs also in Li Dou's *Yangzhou huafang lu* 揚州畫舫錄 [Record of the Painted Barges of Yangzhou] (1795; Yangzhou: Jiangsu Guangling guji keyinshe, 1984), p. 172, appended to his discussion of the various Zheng family gardens.

7. For a short biography of this man, see A. W. Hummel, ed., *Eminent Chinese of the Ch'ing Period, 1644–1912* (Washington: Government Printing Office, 1943) [hereafter, *ECCP*], pp. 148–150. Qian Qianyi seems doubly condemned by history. Despite holding high official rank under the Ming, he was said to have been amongst the first to surrender to the Qing forces when they occupied Nanjing in 1645, offering his services to the new house; by 1647, however, he had been arrested by his new masters under suspicion of Ming-loyalist activities. In a series of edits dating from 1769, the Qianlong emperor (r. 1736–96), whom seems to have harboured a particular and intense animus for Qian, proscribed his entire corpus, fortunately with little lasting effect.

8. In his preface to the collection, dated to the 6th month of 1640 and entitled 'Yaohuang ji xu' 姚黃集 [Preface, *Yao's Yellow Collection*], Qian Qianyi, picking up on issues treated by Ouyang Xiu in his monograph, discusses the extent to which the blooming of Zheng's peony could be constituted

as either an 'Auspicious Flower' (*huarui* 花瑞) or a 'Flower Monstrosity' (*huayao* 花妖), for which, see Qian Zhonglian 錢仲聯, ed., *Qian Muzhai quanji* 錢牧齋全集 [Complete Works of Qian Qianyi] (Shanghai: Shanghai guji chubanshe, 2003), Vol. 3, pp. 885–886. The contemporary bibliophile Huang Shang 黃裳 (b. 1919) owns a copy of this collection that was reprinted in 1762 and from which this preface had later been removed, for which, see his *Qingdai banke yiyu* 清代版刻一隅 [Perspectives on Qing Dynasty Printing] (Jinan: Qi Lu shushe, 1992), pp. 238–239. The collector's seals visible on the facsimile page reproduced in this source reveal that this copy of the collection had once belonged to, amongst others, Ma Yuelu 馬曰璐 (1697–1766?) (on whom, see *ECCP*, pp. 559–560), owner (with his brother Ma Yueguan 馬曰琯, 1638–1755) of another of the great gardens of Yangzhou, the Little Translucent Mountain Lodge, mentioned above.

9. Zheng Yuanxun seems also to have been something of an inveterate joiner of societies; in 1629, for instance, he joined the Fushe 復社 [Restoration Society], established the previous year by Zhang Pu 張溥 (1602–41). For a short biography of this figure, see *ECCP*, pp. 52–53; on the society generally, see William S. Atwell, 'From Education to Politics: The Fu She', in W. T. de Bary, ed., *The Unfolding of Neo-Confucianism* (New York: Columbia University Press, 1975), pp. 333–367. Zheng had previously helped his friend Mao Xiang establish a society in Yangzhou in 1627, another one there in 1636, and in 1642 had sought to revive the fortunes of the Restoration Society at a gathering on Tiger Hill in Suzhou, for which see Zhang Huijian, *Ming Qing Jiangsu wenren nianbiao*, pp. 492, 482, 526 and 564, respectively. Writing sometime before 1630, Dong Qichang 董其昌 (1555–1636), a man who played a crucial role in the creation of Zheng's garden, had claimed that: 'Actually, nine out of every ten who withdrew to the mountains were extravagant men who lived lavishly', for which see Ronald Egan, ed. and trans., *Limited Views: Essays on Ideas and Letters*

by Qian Zhongshu (Cambridge, MA: Harvard University Press, 1998), p. 86.

10. On Gao Jie, see his biography in *ECCP*, pp. 410–411.

11. For a brief account of Gao Jie's role in the complex circumstances of the times, see Lynn A. Struve, *The Southern Ming, 1644–1662* (New Haven & London: Yale University Press, 1984), pp. 23–26.

12. On whom, see the biography in *ECCP*, pp. 701–702.

13. Dai Mingshi, *Dai Mingshi ji* 戴名世集 (Beijing: Zhonghua shuju, 1986), p. 351. With reference to local biographical accounts of Zheng Yuanxun, Li Dou presents a far more nuanced and sympathetic account of Zheng as a man much involved in local activities such as famine relief, for which see Li Dou, *Yangzhou huafang lu*, pp. 169–170. In her masterful treatment of the history of Yangzhou, Antonina Finnane addresses the historiogaphical issues at play here: 'In place of an arrogant, self-serving Yuanxun selling out the city, the local accounts … have a gentry leader of impeccable credentials endeavoring to serve the city, his actions fatally misunderstood by the city mob. Both versions are essentially loyalist, Dai Mingshi's famously so, but the local chroniclers defended the name of a leading member of the local gentry and his family. The different accounts, including Dai's, are premised on popular lore. Dai Mingshi's account is consistent with popular rumor at the time of Zheng Yuanxun's death and held sway because Dai was eventually celebrated as a Chinese martyr. The Yangzhou literati, however, took issue with these rumors and ensured in the wake of the fall of Yangzhou that Zheng Yuanxun would be honored. Yangzhou scholar Jiao Xun (1763–1820) recorded that he had been killed by "evil people". Li Dou, whose potted biographies of Yangzhou people rarely amount to more than a few lines, accorded the Zheng family and its gardens several pages and recorded the entire story of Zheng Yuanxun's death, based on information in the local gazetteers' (*Speaking of Yangzhou: A Chinese City, 1550–1850*, pp. 83–84).

14. For a discussion of the implications of a 'tour' rather than a 'map' of the represented garden, see Craig Clunas, *Fruitful Sites: Garden Culture in Ming Dynasty China* (Durham: Duke University Press, 1996), pp. 139–144. In his *The Afterlife of Gardens* (London: Reaktion, 2004), John Dixon Hunt discusses some of the implications of such descriptions as that given below by Zheng Yuanxun of his garden, intended as it seems to have been to ensure that ' … the layout was accorded a reception worthy of and fully alert to his design invention': 'One important task of every guidebook is to ensure that we miss nothing of importance. But their use also has the effect, even today, of ritualizing a garden visit, although few such publications derive from their creators … or patrons … . Guidebooks lead and instruct, ensuring that the majority of visitors will follow the same route, observe and understand the same things, and so begin to constitute something like a ritual visitation … . Guidebooks are essentially a modern product, responding to the growing taste for tourism and its ritualistic aspects But something equivalent did exist in the Renaissance, and these poetical celebrations and descriptions would have promoted the reputations and prestige of princely and aristocratic owners who clearly wanted their gardens to speak fully and appropriately to visitors as well as those who would read them as armchair travellers; one consequence would be that all visitors who used these texts would share a response' (pp. 154–155).

15. Ji Cheng 計成 (*zi* Wupi 無否, *hao* Foudaoren 否道人, b. 1582), from Wujiang 吳江 in Suzhou Prefecture. For a short biography (by Lienche Tu Fang) of this most famous of Chinese garden designers, see L. Carrington Goodrich and Chaoying Fang, eds, *Dictionary of Ming Biography, 1368–1644* (New York & London: Colombia University Press, 1976), Vol. 1, pp. 215–216. The best modern edition of Ji Cheng's *Yuanye* 園冶 is Chen Zhi, ed., *Yuanye zhushi* 園冶注釋 (Beijing: Zhongguo jianzhu gongye chubanshe, 1981); for a fine translation of this work, see Alison Hardie,

trans., *The Craft of Gardens*.

16. Chen Zhi, ed., *Yuanye zhushi*, pp. 31–35; Alison Hardie, trans., *The Craft of Gardens*, pp. 29–31.

17. Li Dou, in his account of the Garden of Reflections (*Yangzhou huafang lu*, pp. 167–169), although he bases himself on Zheng Yuanxun's own account, makes this dimension of the garden even more explicit when he tells us that the garden was sited on an 'elongated islet' (*changyu* 長嶼) in the midst of South Lake (*Nanhu* 南湖).

18. Elsewhere in his book, Li Dou describes the effect of such conceits: 'Touring this space one feels oneself to be like an ant crawling through the twisting eye of a pearl, or to have encountered a screen of tinted glass, for every new twist and turn leads one on to yet another splendour' (遊其間者如蟻穿九曲珠又如琉璃屏風曲曲引人入勝), for which see his *Yangzhou huafang lu*, p. 139.

19. In his *The Craft of Gardens*, Ji Cheng argues: 'To borrow from the scenery means that although the interior of a garden is distinct from what lies outside it, as long as there is a good view you need not be concerned whether this is close by or far away, whether clear mountains raise their beauty in the distance or a purple-walled temple rises into the sky nearby. Wherever the view within your sight is vulgar, block it off, but where it is beautiful, take advantage of it; never mind if it is just of empty fields, make use of it all as misty background. This is what is known as skill in fitting in with the form of the land. However, even if the form is fitting and the design follows the lie of the land, but the owner still does not get the right person to carry out the work, and in addition is reluctant to spend money when necessary, then any work which may have been done previously will be wasted along with his present efforts'. See, Chen Zhi, ed., *Yuanye zhushi*, pp. 41–42; Alison Hardie, trans., *The Craft of Gardens*, pp. 39–40.

20. All three of Zheng Yuanxun's brothers owned celebrated gardens. Writing in the 1790s, Li Dou notes that the plaque that had once hung above the main gate of the Garden of Reflections had long been lost and even the local gazetteers served to perpetrate some degree of confusion about the actual site of the former garden. All that remained of the garden, he tells us, was a single rock, now inlaid within the door of Old Man Xiao's shop along Buying and Selling Street, for which, see his *Yangzhou huafang lu*, pp. 168–169. In 2004, a plan for the recreation of the Garden of Reflections, on its original site in Yangzhou, was announced; I am unsure what progress has been made in this respect.

21. See Fang Xiangying, 'Chongqi xiuyuan ji' 重葺休園記 [A Record of the Restoration of the Garden of Rest], in Chen Zhi and Zhang Gongchi, eds, *Zhongguo lidai mingyuan ji xuanzhu*, p. 315. Zheng Xiaru (*zi* Shijie 士介, *hao* Sian 俟菴) was Zheng Yuanxun's youngest brother. Li Dou, however, writing sometime before 1795, tells us that after the passage of more than a century the ruins of the Garden of Reflections remained (*yizhi youcun* 遺址猶存) and that just to the south of where the garden had once been stood the shrine dedicated to Zheng Yuanxun and his brother Zheng Yuanhua 鄭元化 (*zi* Zanke 贊可), former master of the Garden of the Fine Trees (Jiashu yuan 嘉樹園).

22. 'Longing and Belonging in Chinese Garden History', in Michel Conan, ed., *Perspectives on Garden History* (Washington, DC: Dumbarton Oaks Research Library and Collection, 1999), pp. 205–219.

23. Chen Jiru 陳繼儒 (*zi* Zhongchun 仲醇; *hao* Meigong 眉公, Meidaoren 麋道人; 1558–1639). For a short biography (by Fang Chaoying) of perhaps the pre-eminent cultural figure of this age, see *ECCP*, pp. 83–84.

24. Stanislaus Fung, trans., in his 'Longing and Belonging in Chinese Garden History', pp. 214–215. For the original text, see Wei Yong, *Bingxue xie* (Shanghai: Zhongyang shudian, 1935), Vol. 1, pp. 1–6.

25. A reference to 'The Turning of Heaven' chapter of the *Zhuangzi* 莊子: 'Benevolence and righteousness are the grass huts of the former kings; you may stop in them for one night but you mustn't tarry there for long' Burton Watson, trans., *The Complete Works of Chuang Tzu* (New York: Columbia University Press, 1968), p. 162.

26. For a suggestive discussion of this general issue, see Pierre Ryckmans, *The Chinese Attitude Towards the Past* (Canberra: The Australian National University, 1986).

27. For a biography of this man, see *ECCP*, pp. 203–205.

28. Quan Zuwang, 'You gu shuibu Zheng jun xiuyuan, yong Xiegu jiuyun' 遊故水部鄭君休園用蠏谷舊韻 [A Tour of the Garden of Rest of the Late Master Zheng of the Ministry of Works, to the Rhyme Formerly used by Ma Yueguan], *Quan Zuwang ji huijiao jizhu* 全祖望集匯校集注 (Shanghai: Guji chubanshe, 2000), Vol. 3, p. 2098.

29. The phrase 'handfuls of stone' (*yiquanshi* 一拳石) occurs in the 'Zhongyong' 中庸 [Doctrine of the Mean]. In a recent translation by Roger T. Ames and David L. Hall, the complete passage reads: 'As for mountains, they are just an accumulation of handfuls of stone, and yet given their expanse and size, grasses and trees grow on them, birds and beasts find their refuge in them, and deposits of precious resources are replete within them', for which, see *Focusing the Familiar: A Translation and Philosophical Interpretation of the Zhongyong* (Honolulu: University of Hawai'i Press, 2001), p. 108.

30. Dong Qichang (*zi* Xuanzai 玄宰; *hao* Sibai 思白, Xiangguang 香光), painter, calligrapher, and pre-eminent art historian. For a short biography (by Fang Chaoying), see *ECCP*, pp. 787–789. Zheng Yuanxun had paid a call on Dong in Songjiang during the winter of 1628, only to find him detained elsewhere, for which see Wang Shiqing 汪世清, 'Dong Qichang de jiaoyou' 董其昌的交游 [Dong Qichang's Circle], in Wai-kam Ho, ed., *The Century of Tung Ch'i-ch'ang, 1555–1636* (Seattle & London: The University of Washington Press, 1992), Vol. 2, p. 481. The previous year, Zheng Yuanxun had included in an anthology that he compiled and published an early statement of Dong's influential theory of the Southern and the Northern Schools of

painting, on which topic, see James Cahill, *The Distant Mountains: Chinese Painting of the Late Ming Dynasty, 1570–1644* (New York: Weatherhill, 1982).

31. Built by Emperor Yang of the Sui dynasty (r. 605–617) once he had established his southern capital at Yangzhou, Labyrinth Tower (Milou 迷樓) came to symbolise both the decadence and extravagance that characterised his reign and its inevitable consequence — dynastic collapse. Emperor Yang was murdered within his tower.

32. The lasting fame of Level with the Mountains Hall (Pingshantang 平山堂) was a result of its association with the great Song dynasty scholar Ouyang Xiu who had once served in Yangzhou and whose monograph on the peony was mentioned earlier. On this site and its restoration during the Qing, see Tobie Meyer-Fong, *Building Culture in Early Qing Yangzhou* (Stanford, CA: Stanford University Press, 2003).

33. In allusion to the iconic Chinese representation of a place of refuge, Tao Yuanming's 陶淵明 (365–427) 'Taohua yuan ji' 桃花源記 [The Peace Blossom Spring], for a translation of which (by J. R. Hightower), see John Minford and Joseph S. M. Lau, eds, *Classical Chinese Literature: An Anthology of Translations: Volume I: From Antiquity to the Tang Dynasty* (New York: Columbia University Press and Hong Kong: The Chinese University Press, 2000), pp. 515–517.

34. The 'yellow' here refers not to the colour of the rocks but to their original geographical provenance, as Ji Cheng makes clear in his *Yuanye*, the relevant section of which reads, in Alison Hardie's translation: 'Huang Rocks [黃石 (yellow rocks)]. Huang rocks are obtainable all over the place. Their texture is firm and does not admit the adze or chisel, and the striations in the stone are rough and coarse. Places such as Huangshan near Changzhou, Yaofengshan near Suzhou, Tushan near Zhenjiang, and all along the Yangtze to above Caishi all produce these rocks. Vulgar people are aware only of their hefty appearance and not of their subtle attraction', see Alison Hardie, trans., *The Craft of Gardens*, p. 116. In a footnote to this section, Hardie adds: 'Although

strictly speaking the term "Huang rocks" should refer to rocks from Huangshan (whichever Huangshan this may mean), it is generally used to refer to rocks which are of a fairly regular shape, with comparatively flat surfaces, by contrast with the fantastic forms of Great Lake rocks' (p. 137).

35. Su Shi 蘇軾 (1037–1101), the great poet of the Song dynasty. For a short biography of this man, see Herbert Franke, ed., *Sung Biographies* (Wiesbaden: Franz Steiner, 1976), Vol. 3, pp. 99–968.

36. For an illustration of this pattern, see Chen Zhi, ed., *Yuanye zhushi*, p. 175.

37. On the first two of these qualities, see John Hay, 'Structure and Aesthetic Criteria in Chinese Rocks and Art', *RES* (1987), 13, pp. 6–22.

38. A reference to a rhapsody so titled ('Zhao yinshi' 招隱士) attributed to a poet at the court of Liu An 劉安 (?179–122 BCE), Prince of Huainan. In the translation of David Hawkes, the first lines of this poem go: 'The cassia trees grow thick/In the mountain's recesses,/Twisting and snaking,/Their branches interlacing./The mountain mists are high,/The rocks are steep./In the sheer ravines/The waters' waves run deep.', for which see David Hawkes, trans., *The Songs of the South: An Ancient Chinese Anthology of Poems by Qu Yuan and Other Poets* (Harmondsworth, UK: Penguin, 1985), p. 244.

39. Ni Yuanlu 倪元璐 (*zi* Yuru 玉汝; *hao* Hongbao 鴻寶; 1594–1644). For a short biography (by George A. Kennedy), see *ECCP*, p. 587.

40. The conceit of this name is impossible to convey fully in translation. To the character *mei* 眉 (eyebrow), Zheng adds the 'Water Radical', giving *mei* 湄 (the margin of a lake). In an architectural context, the character *rong* 榮 (glory, splendour, etc.) gives the technical meaning of 'overhanging eaves'.

41. Chen's style name (Meigong 眉公) translates as 'Duke Eyebrow', hence the name 'Drenched Eyebrow's Prominence' above. This belvedere lent its name to a collection of occasional prose of the Ming dynasty that Zheng Yuanxun compiled with Chen Jiru's help sometime in the early 1630s: *Meiyouge wenyu* 媚幽閣文娛 [Literary Amusements

from the Belvedere of the Love of Solitude]. In his 'Preface' to this work, Chen describes Zheng as: 'a true hero, both upright and fearless' (*leiluo xia zhangfu* 磊落俠丈夫), for which see Hu Shaotang 胡紹棠, ed. *Chen Meigong xiaopin* 陳眉公小品 [Essays of Chen Jiru] (Beijing: Wenhua yishu chubanshe, 1996), p. 25.

42. Li Bo 李白 (701–62), the great Tang dynasty poet. This line comes from his poem entitled: 'Xunyang zijijigong gan qiu zuo' 尋陽紫極宮感秋作 [Written When Moved by the Autumn in the Purple Extremity Palace of Xunyang], for which see *Li Taibai quanji* 李太白全集 [The Complete Works of Li Bo] (Beijing: Zhonghua shuju, 1977), Vol. 2, p. 1114.

43. Wang Chun 王醇 (*zi* Xianmin 先民; d. 1627), from Yangzhou. See *Mingren zhuanji ziliao suoyin* 明人傳記資料索引 [An Index to Ming Dynasty Biographical Sources] (Taibei, 1965–66), Vol. 1, p. 69.

44. Ji Cheng begins his *The Craft of Gardens* by addressing the issue that Zheng Yuanxun alludes to here: 'Generally, in construction, responsibility is given to a "master" who assembles a team of craftsmen; for is there not a proverb that though three-tenths of the work is the workmen's, seven-tenths is the master's? By "master" here I do not mean the owner of the property, but the man who is master of his craft'. See, Chen Zhi, ed., *Yuanye zhushi*, p. 41; Alison Hardie, trans., *The Craft of Gardens*, p. 39.

45. This is perhaps a veiled allusion to the most famous of 'Yangzhou dreams', that of the late Tang dynasty poet Du Mu 杜牧 (803–852) in his poem 'Qianhuai' 遣懷 [Easing My Heart] which, in A. C. Graham's translation, reads: 'By river and lakes at odds with life I journeyed, wine my freight:/Slim waists of Chu broke my heart, light bodies danced into my palm./Ten years late I wake at last out of my Yangzhou dream/With nothing but the name of a drifter in the blue houses', for which see John Minford and Joseph S. M. Lau, eds, *Classical Chinese Literature: An Anthology of Translations: Volume I: From Antiquity to the Tang Dynasty*, p. 915.

46. Hanjiang 邗江 is an archaic name for Yangzhou.

On loanwords and calques: *where the language of design meets the language of geology*

JACKY BOWRING

Introduction

The flows of garden-design models around the globe create a dynamic field in which imported approaches interact with local settings in a range of ways. Such interactions embody the play of power in different ways, and as Stephen V. Ward notes of planning models there are perspectives from 'importing' and 'exporting' nations.[1] There are also dynamic exchanges between the imported models and the local setting into which they are introduced. This local setting might be wholly erased to accommodate the imported model, or there may be something of an exchange between the two, where elements may be retained or added. This paper explores the encounter of garden-design models and local settings at the level of individual elements, focusing in particular on geological features in gardens.

Gardens are often referred to as having a 'design language', a term which immediately evokes ideas of conversations, quotations, and translations, when considering the interactions between an imported model and the local setting.[2] Design language is made up of conventions, formulae, and frameworks that structure space and determine aesthetic appeal. Nature, too, has a language, with vocabularies of botanical and geological elements, constituting a syntax that reflects biophysical regions.

Design language can be very insistent, bringing its modes of composition to bear on whatever landscape is encountered. This place of meeting is an intensely creative zone, where the negotiation becomes crystallized into new linguistic formations. The creative fusion produces the languages of pidgins and creoles, hybrid languages which embed the vocabulary from both source languages to form new and creative ways of communicating. And at the scale of the words themselves, loanwords and *calques* can develop — the borrowing of some words, and the tracing of others. This metaphorical frame for the apprehension of the encounter between design language and the language of a natural setting offers a framework for critical analysis. While such a linguistic frame might be presumed to be a product of post-structuralism, which has accounted for the linguistic turn in many disciplines,[3] for gardens, language is an ancient and engrained analogy, as outlined below. The implied question of such an analogy is, of what use is it? The connections made here among gardens and loanwords and *calques* argue for a method of reading gardens as an analytical approach for garden historians when used retrospectively. There is potential, too, for such a method to be used in the creation of gardens, as means of critically considering how site features might be enhanced, altered, or even created.

In looking specifically at the incorporation of geological elements, such as glacial traces and dunes, the intention is to highlight the importance of critically considering aspects of a garden which might be taken as natural. In some cases they may be natural, but they may also be augmented, or even artificial. Casting these geological elements as loanwords or *calques* offers insights into how this process of linguistic migration might occur at the scale of the components of a garden. Rather than the phrase-scale of a quotation or paraphrase, as in David Leatherbarrow, John Dixon Hunt and Laurie Olin's lexicon,[4] loanwords and *calques* operate at the level below this, offering the potential for a 'textual analysis' of a garden through critical consideration of its individual elements, its 'words'.

'Nature is a language, can't you read?'[5]

The metaphorical synthesis of nature and language is deeply embedded in Western culture, and establishes the idea of a 'natural language' of a setting which interacts with the imported language of design. This relationship between nature and language can be traced through a centuries-long legacy, as in William J. Mills' description of how language, in the form of a book, became a pervasive landscape metaphor, so that: 'By the Middle Ages we find that the book of nature has become adopted universally as the image through which the environment is to be understood.'[6] The 'book of nature' embodied the ideas of an author or creator, and carried with it ideas of stewardship, where 'The world's a book in folio, printed all/With God's great works in letters capital:/Each creature is a page; and each effect/A fair character, void of all defect.'[7] This is echoed in the Doctrine of Signatures, with language evoked through the forms of the elements that make up the environment, where 'the finger of God hath left an inscription upon all His works, not graphical or composed of letters, but of their several formes, constitutions, parts, and operations, which aptly joined together do make one word that doth express their natures'.[8]

Metaphorical connections between language and landscape continue throughout Western literature, as in the passage from Shakespeare's *As You Like It* (1599), where the Forest of Arden is revealed as a rich resource, and Duke Senior describes how he, 'Finds tongues in trees, books in the running brooks,/Sermons in stones, and good in every thing [sic].'[9] And, in Novalis' novel, *The Novices at Sais* (1798), one of the travellers refers to nature's own language, 'that great cipher which we discern written everywhere, in wings, egg-shells, clouds and snow, in crystals and in stone formations, on ice-covered waters, on the inside and outside of mountains, plants beasts and men, in the lights of heaven … '.[10] This vision of a language of nature is echoed in Ralph Waldo Emerson's *Nature* (1836), wherein, 'A life in harmony with nature, the love of truth and of virtue, will purge the eyes to understand her text. By degrees we may come to know the primitive sense of the permanent objects of

nature, so that the world shall be to us an open book, and every form significant of its hidden life and final cause.'[11] The enduring desire for legibility in the natural world is pervasive, from the ancient Babylonians' vision of stars as the writing of the sky, to Marcus Clarke's belief that: 'In Australia alone is to be found the Grotesque, the Weird, the strange scribblings of nature learning how to write.'[12]

A language of context can therefore be seen in the context of nature, with its own coherent structure, rules and meaning. And, for landscape architecture, the idea of a linguistic frame for the environment is naturalized within the common frames of reading and writing the site, as in the specific context of geology, where Elizabeth Meyer refers to the ancient landforms of the Prospect Park area as the 'formal language of glacial geomorphology', and explains how the site plan that depicts the topography 'allows the ground to speak'.[13] Anne Whiston Spirn's *The Language of Landscape* circles around the idea of metaphorical connections between the workings of language and those of landscape.[14] For Spirn, this language is not simply a means of framing landscape, but of making meaningful and responsible decisions about how to intervene in the natural world.

Pidgins and creoles

Languages of contact are arenas for both tensions and creative hybridization. A pidgin is a language which forms 'when two mutually unintelligible speech communities attempt to communicate',[15] and it is characterized by 'a limited vocabulary, reduced grammatical structure, and a much narrower range of functions, compared to the languages which gave rise to them'.[16] Once a pidgin is fully adopted and becomes the mother tongue of a community, it is known as a creole, bringing with it 'a major expansion in the structural linguistic resources available — especially in vocabulary, grammar and style'.[17]

In the same ways that language itself responds to the dynamics of contact, so too does the language of design. Encountering an unfamiliar culture or an alien nature pushes a design language to adjust and adapt, adopting elements of this new setting. Just as words or phrases cross between languages, creoles and pidgins can form where a design language interacts with an unfamiliar context. This language of encounter is epitomized by the arrival of the Picturesque in New Zealand, bringing a highly developed compositional system into contact with a strange new landscape, producing what can be called a Pidgin Picturesque.[18]

The Picturesque was a particularly codified design language, described as a 'codex',[19] a 'formula'[20] or even a recipe as in the 'art of cooking nature'.[21] Contact between the Picturesque and New Zealand nature was initially one of mutually unintelligible speech communities attempting to communicate. The Picturesque had grown out of the British landscape, inflected with the experience of the Continent, and with this the contextual language of a particular ecology, climate and geomorphology. New Zealand's landscape did not speak this language. While there were efforts to mould the Picturesque to the new landscape, some early responses illustrated the challenge of the language of contact. Lord Lyttelton pointed out the lack of the makings of the Picturesque on the vast, flat Canterbury Plains which abut the city of Christchurch, suggesting this landscape could be described as 'repulsive'.[22] The vegetation, too, was problematic, and artists often translated what they saw into a familiar visual language, such as Charles Heaphy's painting of Taranaki Mt Egmont, where the vegetation is limned in a delicate European manner which is at odds with the solidity of New Zealand bush (figure 1).

However, while painters could 'translate' what they saw into their familiar visual language, for the physical reality of garden design this was more difficult. One early settler described the problem as the 'certain stiffness in the appearance of a New Zealand forest, which contrasts unfavourably with the fresh tender green of an English wood'.[23] A pidgin emerged when this imposition gave way to crossbreeding of languages, the language of design and the language of context both yielding a little until a new form emerges. The indigenous vegetation was coerced into performing compositional tasks suited to the Picturesque, and gardens developed around the necessary topographical substrates such as river boundaries or gentle hills. In Christchurch, for example, despite Lord Lyttelton's derogatory comment regarding the repulsiveness of the landscape setting, Deans' Bush next to the river in Riccarton was described as 'a spot with which lovers of the Picturesque must be pleased'.[24]

The relationship between the Picturesque and the local eventually became engrained and familiar, passing into a creole, a fluent form in its own right. This creolized Picturesque can be seen in some of New Zealand's 'native gardens' of the 1960s and 1970s. These gardens maintained the artifice of nature that underlies the Picturesque, reducing the compositional rules to domestic scale, and utilizing indigenous plants. This hybrid of exotic and

FIGURE 1. *Charles Heaphy, 1820–1881, Mt Egmont from the southward. [September? 1840]. Source: Alexander Turnbull Library, Wellington, New Zealand, Ref: C-025-008. © Alexander Turnbull Library. Reproduced by permission of Alexander Turnbull Library. Permission to reuse must be obtained from the rightsholder.*

local epitomizes the transformation from pidgin to creole, with the landscape echoing the same kinds of exchange and adaptation as in language.

Loanwords and *calques*

The creative tension of imported design models and local contexts is exemplified by the fusion languages of pidgins and creoles. Leatherbarrow *et al.* conceptualize the dynamics of 'foreign' models through a range of terms including translation, imitation and adaptation, all of which adopt the linguistic analogy.[25] Two of their terms resonate strongly at the level of individual elements, as in the case of the geological features which I explore in this paper; paraphrase and quotation. They summarize the difference between these two operations as, 'Quotations repeat, paraphrases restate'.[26] In the same way that one is a direct repetition and the other is a restatement, loanwords and *calques* differ in terms of the way this adoption occurs. While both loanwords and *calques* contribute to the formation of a language of fusion, they do so in different ways, and this suggests some intriguing parallels in considering the transformation of language as a metaphor for the contact between design languages and their new contexts.

Loanwords are appropriated from one language into another, and in English there are many examples, such as words loaned from Italian (*pizza, plaza*) and from French (*chef, fierce*). A loanword retains its original form and meaning, and is swallowed whole by the other language, eventually becoming completely integrated. Yet on close scrutiny these words can be seen as 'other' to the dominant language.

A *calque* is literally a 'tracing', and involves taking a literal translation of a word or phrase from the encounter language, such as the term 'blue blood', which is a translation from the Spanish *sangre azul*. The original term referred to nobles being identified by the clarity of their blue veins in translucent white skin, as evidence that they were not contaminated by other races, or that they were not of a low class and would be tanned from outdoor labouring. Another example is 'flea market', a *calque* of *marché aux puces*, which reputedly referred to the fleas in the items for sale. These *calques* retain the original sense of the term but are linguistically reworked — they tend to lose the integrity of the source term as they have been translated in small pieces. As Miriam Meyerhoff explains: 'A *calque* can be defined quite specifically in linguistic terms, and in the field of language contact, a *calque* is usually used to refer to the direct translation, morpheme by morpheme, or word for word, of concepts and syntactic structures that originated in one language and can be shown to be (or be argued to have been) a historical introduction into another.'[27]

Calques can take a very specific situation and generalize it through the process of translation, such as the phrase *Devil's advocate*, a *calque* of the term *advocātus diabolī*. While the original term referred to a specific situation relating to the Catholic Church where an appointed official presents arguments against a proposed canonization or beatification, the *calque* subsequently applied to any situation where someone puts up an argument for the sake of highlighting a specific point, even if they do not believe it.

In a designed landscape, a loanword is the inclusion of an element of the existing context into a composition. It retains its otherness in physical form, yet becomes very familiar, such that it may even become naturalized and viewers do not realise it is from another 'language'. Just as words like pizza and plaza are so linguistically entrenched that they are exist seamlessly in English, so too are landscape loanwords.

A landscape *calque* is traced from another 'language', rather than being adopted as a given. A *calque* is an artificial construction, translating a preexisting element into the new landscape. The *calque* masquerades with a sense of being a natural component of the landscape, but is really an invented element. And like a linguistic *calque* it can take a specific situation from its context and generalize it to the wider environment.

Landscape loanwords

Landscape loanwords appear in nineteenth-century gardens, paralleling the ways in which Picturesque gardens in England had appropriated existing landscape features, such as cliffs, hills and rivers. Jay Appleton drew attention to the close relationship between topography and the Picturesque in Britain, and offered the thesis that the 'aesthetic criteria of the Picturesque made very special demands on the geological dimensions of the *genius loci*'.[28] Appleton contended that, 'with few exceptions, the opportunities afforded by nature for the achievement of those effects which were admired and advocated by the writers on picturesque landscape were unevenly distributed. The vast majority of *natural* features conducive to picturesque design — cliffs, waterfalls, ravines, precipices, etc. — occur among the Palaeozoic, igneous or metamorphic rocks of Upland Britain ... '.[29]

This elevation of a site's native language into that of the Picturesque is echoed in the USA by Frederick Law Olmsted and Calvert Vaux. Elizabeth Meyer draws attention to how Olmsted and Vaux's projects at Prospect Park in Brooklyn (1868) and Central Park in Manhattan (1853) emphasize and exhibit remnant glacial features in the landscape. Rather than simply imposing the eighteenth-century theories of Uvedale Price, William Gilpin and Richard Whately, Meyer suggests that Olmsted and Vaux interpreted them as a need to *read* a site.[30] This meant tuning into the physical structure of the site, and working with the elements — the 'vocabulary' — of the place. However, they did not simply adopt the natural landscape as a totality, and instead went through a process analogous to editing, essentially underlining existing geological elements. Meyer points to how the different aspects of the glacial landscape were enlisted into the design for Prospect Park, establishing an 'armature' which was 'a structure that Vaux and Olmsted reinforced through the disposition of plants and promenades around and across these distinct landforms. Long Meadow, the Ravine and Lookout Hill, and Prospect Lake are characterized as the Beautiful (or Pastoral) and the Picturesque, but they are configured by the uneven advances and retreat of the Wisconsinan glacier across this terrain'.[31]

Although incorporated into the language of the Picturesque, the vocabulary of the glacial landscape retained its form and meaning. In fact, Olmsted and Vaux sought to extend the boundaries of Prospect Park in order to incorporate a wider sweep of landscape types and enhance the integrity of the glacial remnants.[32] Meyer further notes how at Central Park these remnant elements were articulated by the designers in a way which transformed the rocks from being merely circumstantial into being 'significant figural events' (figure 2).[33] With their access to geological cross-sections, they were able to carefully select where to expose the rock through 'subtracting' from the site, and revealing its geological underpinnings. Meyer describes how 'framing and foregrounding the outcrop also highlighted the striations of its surface, formed by glacial grinding. A touch of the sublime was injected into the landscape through both the outcrop's actual size and its grain, which registered the processes of its formation'.[34]

Meanwhile, in Christchurch in 1863 the Botanic Gardens became established, as the city itself moved into its second decade. Here the language of context — the nature that was encountered by the colonizers — was a wild flat landscape of tussocks and ferns. The Avon River meandered through the area that is now the Botanic Gardens, and the landscape was dotted with swamps filled with rushes and flax. Around 7000 years ago the coastline ran through what is now central Christchurch and the central suburb of Riccarton, with the previous traces of sandhills and hollows now largely erased from the urban fabric. Much of the sand and gravel was used for building roads in the expanding city.

When encountering the Pine Mound in the Christchurch Botanic Gardens, it is a revelation to recognize it as part of a long-distant geomorphological process (figure 3). This ancient sand dune was once lapped by the Pacific Ocean, which now lies some 11 kilometres to the east. The sound of the wind in the pines carries an echo of this vanished coastal landscape. Maritime pines (*Pinus pinaster*) were planted on the sand dune in 1871, and the nearly 150-year-old trees' soaring forms create one of the Gardens' sublime moments. Just

FIGURE 2. *A landscape loanword — part of the glacial landscape of the Manhattan Formation is evident in Central Park. Source: photograph by author.*

as Olmsted and Vaux used the remnant glacial geomorphological features as an 'armature' for their work at Prospect Park and Central Park, the Pine Mound is an amplification of a natural geological feature to the status of designed landscape. Meyer noted how Olmsted and Vaux 'supplemented' the eskers at Prospect Park and the ground becomes 'figured' with the 'site emerg[ing] from the frame of landscape'.[35] The process of supplementation and figuration is echoed at the Pine Mound, with the tall pines accentuating the landscape feature of the sand dune. Without the pines, the vertical impact of the mound would be minimal, but their presence acts as an emphasis — an underlining of the loanword. While in a more varied terrain such an element might be overlooked, in the flatness of Christchurch the pine-topped mound becomes a highpoint, topographically and experientially.

Landscape *calques*

Landscape *calques* also appear in nineteenth-century public parks and gardens, particularly in the eclectic gardens of the late nineteenth century, resonating with the Victorian mania for faux nature, collecting and cabinets of curiosities. Landscape *calques* mimicked natural features, translating elements from another landscape language into physical form. Battersea Park was one such example, developed in the late 1860s, on a flat and featureless site on the south bank of the River Thames in London. It featured fake stone work that at a glance looks convincingly authentic (figure 4). Transcending the apparent lack of natural topography, a six hectare lake was created, complete with bays and peninsulas, and an entire geological system was invented through the use of

FIGURE 3. *The Pine Mound, Christchurch Botanic Gardens. Source: photograph by author.*

FIGURE 4. *A landscape* calque — *Pulhamite rockwork at Battersea Park, London. Source: photograph by author.*

Pulhamite stone work. The faux geology exemplifies the operation of a *calque*, or 'trace', of a natural feature, and the artificial rocky crags are a key element of the invented landscape.

Pulhamite was developed by James Pulham, and was a technique for making fake rock, using cement and aggregates to create replicas of various geological elements such as the Millstone Grit of the Pennines.[36] Over 170 Pulhamite constructions were created around the UK, and geology professor Matthew Bennett has studied Pulhamite constructions, observing that those at Battersea Park are 'dramatic and demonstrates [sic] both geomorphological and geological features with a surprising degree of complexity'.[37] This includes aspects, such as jointing and bedding, and the use of graded aggregates to echo real geological structures (figure 5). However, in the same way that careful scrutiny of a linguistic *calque* reveals its constructed nature, the artificiality of Battersea Park's Pulhamite is becoming apparent with age (figure 6).

FIGURE 6. *The base of Pulhamite rockwork showing how the ground is pulling away to reveal the artifice of the fake geology. Source: photograph by author.*

FIGURE 5. *Pulhamite rockwork showing the fake jointing patterns. Source: photograph by author.*

The Village of Yorkville Park in Toronto, Canada, also demonstrates the concept of a landscape *calque*. The park was established in 1993, designed by Martha Schwartz, Ken Smith and David Meyer who won a design competition. The design responds to the park's Victorian context, in an area of late nineteenth- and early twentieth-century terraced housing. This residue of the Victorian era creates a fine grain with the intricate scale of small lots remaining from the terraced houses, lending a sense of compartmentalization to the context. Combined with references to the Victorians' passion for classifying and displaying scientific specimens, this fine grain scale was reflected in the park's box-like structure, which traces the pre-existing lot boundaries from 1878. Each box contains a symbolized tableau of the surrounding landscape, as though in a Victorian display. As a landscape version of a cabinet of curiosities, or a butterfly collection, the Village of

Yorkville Park becomes a microcosm of the vast landscape beyond. However, while the Pulhamite constructions in Victorian England were geological fantasies constructed from artificial rocks, at the Village of Yorkville actual material from the nearby landscape is used to form a geological copy. The rock feature appears to be emerging out of the ground, in the same way as the real glacial remnants do in Prospect Park and Central Park. However, this rock element is a 650-tonne piece of billion-year-old granite cut from the ancient Canadian Shield formation, brought to the park in pieces and reconstructed on site.

Another example of a geological *calque* is Michael van Valkenburgh's Teardrop Park in Manhattan (figure 7), where the striated forms were 'manufactured in an upstate quarry and then reassembled in the park'.[38] Like the Pulhamite of Victorian England and the microcosmic granite dome of Yorkville Park, the rock formations of Teardrop Park are geological fantasies, even

FIGURE 8. *Sign identifying the Teardrop Park rockwork as 'geological sections'. Source: photograph by author.*

though they appear natural, like the geological outcrops in Central Park which is just several stones' throws away. While at Central Park the geology is edited through a process of underlining and amplifying, at Teardrop Park the features (clearly labelled 'Geological Sections' in an exhortation not to climb them; figure 8) are traced from elsewhere and introduced to the site. Central Park's glacial loanwords and Teardrop Park's *calqued* geological sections both appear natural at a glance, yet exhibit different strategies on the part of their designers.

Conclusion

The encounter between a design language and a language of context inevitably involves negotiation, substitution, and a sense of yielding or imposition. The conversation sees the development of new relationships. Pidgins and creoles are created as a consequence of this encounter, in the midst of an almost

FIGURE 7. *Constructed rockwork at Teardrop Park, Manhattan, designed by Michael Van Valkenburgh. Source: photograph by author.*

metamorphic process of heat and pressure. And within this are the elements of vocabulary that sit within the broader matrix of language. In some cases, they are taken directly from the local context into the language of designed landscape as loanwords, while others are tracings or *calques*.

This metaphorical relationship reveals how the existing landscape and the arrival of a designed language undergo a dynamic evolution. The focus on geological elements illustrates how assumed expressions of the natural landscape can say much about the design frame in which a park was created. The large city parks of Prospect Park and Central Park, along with Christchurch's Botanic Gardens, manifest the articulation of geological remnants, emphasized and exhibited within the composition. In contrast, the gardenesque design of Battersea Park, the postmodern Village of Yorkville Park, and Manhattan's Teardrop Park, see an ersatz or perhaps even kitsch invention of pieces of geology.

Beyond these examples it could be possible to explore other way in which geology 'speaks' in the language of design. Even at the planning scale, Anita Berrizbeitia describes how in Boston's Metropolitan Park system Charles Eliot was also working with the underlying geological language to inform the open space network.[39] For Eliot, geology was also a 'language'[40] and a part of his vision of creating a city that was of its place, rather than being defined by politics. And, at the other extreme, Carmen Perrin's cleaned boulder on the Swiss Way[41] or Roxy Paine's *Erratic*[42] exhibit rocks in ways that are once natural (the rock itself) and artificial (the rock's surface treatment and its location). Both Perrin and Paine's rock-works are, interestingly, erratics, rocks which are geologically out of place as a consequence of the processes of glaciation. These amplifications and creations of physical features work in different ways in terms of design — defamiliarizing a taken-for-granted feature, or evoking process.

Both landscape loanwords and landscape *calques* can be taken for natural in the landscape, just as in language the vocabulary which has been naturalized into English does not stand out as being alien. However, close scrutiny of loanwords and *calques* reveals their different subtleties. And, like the example of devil's advocate being a *calque* which generalizes a specific situation, the Pulhamite rocks of Battersea Park create a generic idea of geology that relocates specific rock formations into the middle of a very different setting. The Pulhamite rocks have become visually naturalized, settling into the surrounding park landscape, just as *calqued* words are absorbed into language. As part of the wide sweep of gardens and designed landscapes, these geological examples illustrate the potential of language as a frame for exploring the dynamism of the contact zone between design and its setting.

Acknowledgements

Thanks to James Beattie and the anonymous reviewers for their constructive feedback and extremely helpful suggestions.

Disclosure statement

No potential conflict of interest was reported by the author.

NOTES

1. Stephen V. Ward, 'The International Diffusion of Planning: A Review and a Canadian Case Study', *International Planning Studies*, iv/1, 1999, p. 58.
2. David Leatherbarrow, John Dixon Hunt and Laurie Olin, 'Some Terms for the Transposition of Gardens Between Countries', *Studies in the History of Gardens & Designed Landscapes: An International Quarterly*, xxxi/4, 2011, pp. 355–356.
3. For example, Trevor J. Barnes and James S. Duncan (eds), *Writing Worlds: Discourse, Text and Metaphor in the Representation of Landscape* (London; New York: Routledge, 1992); Stephen Bending, 'Re-reading the Eighteenth-century English Landscape Garden', *The Huntington Library Quarterly*, lv/3, 1992, pp. 379–399; James S. Duncan, *The City as Text: The Politics of Landscape Interpretation in the Kandyan Kingdom* (Cambridge: Cambridge University Press, 1990).

4. Leatherbarrow, Dixon Hunt and Olin, 'Some terms'.

5. From the song, 'Ask' on the album *Rank*, by The Smiths (Rough Trade Records, 1988).

6. William J. Mills, 'Metaphorical Vision: Changes in Western Attitudes to the Environment', *Annals of the Association of American Geographers*, lxxii/2, 1982, p. 239.

7. Mills, 'Metaphorical Vision', p. 239.

8. Mills, 'Metaphorical Vision', p. 240.

9. William Shakespeare, *As You Like It*, 1599, Act II, Scene I, cited in Adrian Poole, *Shakespeare and The Victorians* (London: A&C Black, 2014), p. 130.

10. Novalis (Georg Philipp Friedrich Freiherr von Hardenberg), *The Novices at Sais*, 1798, cited in Andrew Cunningham and Nicholas Jardine, *Romanticism and the Sciences* (Cambridge: Cambridge University Press, 1990), p. 6.

11. Ralph Waldo Emerson, *Nature*, 1836, in David Greenham (ed.), *Emerson's Transatlantic Romanticism* (New York: Palgrave Macmillan, 2012), p. 51.

12. Marcus Clarke, 'Preface', in Adam Lindsay Gordon, *The Poetic Works of Adam Lindsay Gordon* (Raleigh, NC: Hayes Barton Press, 1955), p. 7.

13. Elizabeth K. Meyer, 'The Expanded Field of Landscape Architecture', in George F. Thompson and Frederick R. Steiner (eds), *Ecological Design and Planning* (New York: Wiley, 1997), p. 56.

14. Anne Whiston Spirn, *The Language of Landscape* (New Haven: Yale, 2000).

15. David Crystal, *An Encyclopedic Dictionary of Language and Languages* (Oxford: Basil Blackwell, 1992), p. 302.

16. David Crystal, *The Cambridge Encyclopedia of Language* (Cambridge: Cambridge University Press, 1987), p. 334.

17. Crystal, *The Cambridge Encyclopedia*, p. 336.

18. Jacky Bowring, 'Pidgin Picturesque', *Landscape Review*, ii, 1995, pp. 56–64.

19. Jack Calow, 'Fine Art and the Picturesque', *The Picturesque: Being the Journal of the Picturesque Society*, 1, 1992–1993, p. 20.

20. Carl Paul Barbier, *William Gilpin: His Drawings, Teaching and Theory of the Picturesque* (Oxford: Oxford University Press, 1963), p. 99.

21. Peter Bicknell, *Beauty, Horror and Immensity: Picturesque Landscape in Britain* (Cambridge: Cambridge University Press, 1981), p. ix.

22. Cited in Thelma Strongman, *The Gardens of Canterbury: A History* (Wellington: Reed, 1984), p. 33.

23. Cited in Trudie McNaughton (ed.), *Countless Signs: The New Zealand Landscape in Literature* (Auckland: Reed Methuen, 1986), p. 203.

24. C. Warren Adams, *A Spring in the Canterbury Settlement* (London: Longman, Brown, Green, 1853), p. 33.

25. Leatherbarrow, Dixon Hunt and Olin, 'Some terms'.

26. Leatherbarrow, Dixon Hunt and Olin, 'Some terms', p. 355.

27. Miriam Meyerhof, 'Replication, Transfer, and Calquing: Using Variation as a Tool in the Study of Language Contact', *Language Variation and Change*, xxi/3, 2009, p. 298.

28. Jay Appleton, 'Some Thoughts on the Geology of the Picturesque', *Journal of Garden History*, vi/3, 1986, p. 271.

29. Appleton, 'Some thoughts', p. 287.

30. Elizabeth Meyer, 'Site Citations: The Grounds of Modern Landscape Architecture', in Carol J. Burns and Andrea Kahn (eds), *Site Matters: Design Concepts, Histories, and Strategies* (New York: Routledge, 2005), pp. 96–97.

31. Meyer, 'Site Citations', p. 56.

32. Meyer, 'Site Citations', pp. 54–55.

33. Meyer, 'Site Citations', p. 108.

34. Meyer, 'Site Citations', p. 108.

35. Meyer, 'Site Citations', p. 56.

36. Matthew R. Bennett, *Geology on Your Doorstep: The Role of Urban Geology in Earth Heritage Conservation* (London: Geological Society of London, 1996).

37. Bennett, *Geology on your Doorstep*, p. 81.

38. Matt Dallos, 'Seeing Landscape: Geography, Autobiography, and Metaphor', *Studies in the History of Gardens and Designed Landscapes*, xxxiv/2, 2014, p. 147.

39. Anita Berrizbeitia, 'Between Deep and Ephemeral Time: Representations of Geology and Temporality in Charles Eliot's Metropolitan Park System, Boston (1892–1893)', *Studies in the History of Gardens and Designed Landscapes*, xxxiv/1, 2014, pp. 38–51.

40. Berrizbeitia, 'Between Deep and Ephemeral Time', p. 38.

41. See Georges Descombes, 'Shifting Sites', in James Corner (ed.), *Recovering Landscapes: Essays in Contemporary Landscape Architecture* (New York: Princeton Architectural Press, 1991), p. 84.

42. See, for example, the Public Art in LA website entry http://www.publicartinla.com/sculptures/erratic.html.

Gardens, history and the designer: contributions to historiography

IAN HENDERSON

Introduction

In accounts of garden history, the garden itself, its physical attributes and spatial organization, is often strangely absent. The focus instead often appears to be on attributing meaning derived from literary and artistic references or historical precedents. I argue that the primary text is the garden itself, and that a failure to account for its physicality — where it exists or has existed in three dimensions — deprives the historian of potential analytical and interpretive opportunities. This paper argues for a place in garden history for analysis of the kind that landscape designers engage in, derived from the design process, and involving both site analysis and design analysis.

The *how* of meaning

Rod Barnett in *Landscape Review*[1] claims that historians generally have viewed gardens of the modern period[2] to be bereft of meaning.[3] He claims that they have failed to recognize the meaning or type of meaning sought by the designers of the period. Barnett's paper was a response to a number of articles on gardens and meaning: notably, centred around a debate on whether gardens have meaning, must have meaning, or even can have meaning.[4] I am not suggesting that gardens are without meaning or that they exist in isolation from literary references or historical precedents. Like any other artefact, they reside within a rich cultural context subject to many diverse, complex, and ambiguous or hybrid cultural interpretations, often of a specific, even personal nature. Rebecca Solnit, in her essay, 'The Colour of Shadows, the Weight of Breath, the Sound of Dust', attributes great significance to the associations that everyday objects have for memory and the capacity to transport us to different times and places.[5] According to Solnit, such associations uplift the object to a much higher value than its apparent intrinsic one. She argues that the object is often of little value itself compared to the associations that accompany it. So, such associations, at whatever scale or in whatever derivations, whether personal or of a broader cultural origin, are not irrelevant in the search for meaning in objects.

Like Solnit's everyday objects, gardens, too, contain the cultural ideas of the societies from which they emerge. Ideas of nature and especially its idealization are many and varied, including that most primal one identified by Clarence Glacken in his seminal work, *Traces on the Rhodian Shore*:

> The pleasure garden … was often conceived of as a simulation of the Garden of Eden. To cultivate the garden was more than a task for the holy; it was also a reliving of part of the creation. The experience was aesthetic and religious.[6]

The very title of John Dixon Hunt's *Greater Perfections* also alludes to such notions of idealization through reference to the ideas of natural historian Francis Bacon.[7] For Barnett, gardens are rich in allegory, symbolism, metaphors, utopias and heterotopias: for him, many statements in the literature on gardens 'attest to the garden as a locus of cultural, metaphysical and spiritual meaning'.[8] Specific examples of meaning acquired through association might include Chinese scholar gardens, which exist in a tight interdependent cluster of allied cultural artefacts, including painting, poetry and calligraphy, besides particular natural environments, especially mountain peaks.[9] These various expressions cross-fertilize and inform each other to construct a comprehensive whole. In like manner, Japanese *karesansui* (dry raked) gardens are often associated with Zen Buddhist temples.[10] All such associations expand the meaningfulness of these gardens, giving them a richness within their own cultural milieu, of both place and time, though it is most likely that any depth of meaning is selectively created, directed at, and understood by a literate elite. However, importantly, these gardens can still be appreciated today, even by people from a different culture, without, in the former example, necessarily having much knowledge of, say, the paintings, poetry, and calligraphy of Ming-dynasty (1368–1644) Suzhou literati, or in the latter, knowing Buddhist mantras. This appreciation derives through the experience of the physicality of the garden itself.

Hunt suggests that garden histories must include 'the melding of palpable (phenomena) with impalpable (noumena)'.[11] Their melding is probably necessary to complete the stories of any garden, but an apprehension of phenomena is necessitated by initial experience, which he also recommends as taking place prior to familiarity with its histories.[12] Marc Treib, in his essay, 'Sources of Significance; the Garden in Our Time', writes that: '[t]hroughout history, the garden has served two primary purposes: as a zone of modulated and intensified sensual experience; and as a vehicle for expressing symbolic, political, and religious ideas beyond the realm of its tangible materials'.[13] Both these purposes provide the analyst of any garden with much to discuss. It is in the former — the 'modulated and intensified sensual experience' — that the designer may go beyond mere description to find significance and meaning. This paper attempts to discover those areas where both sensual experience and designer knowledge intervene. It then employs a visit to a significant Modernist garden to demonstrate this meeting of sensual experience with intangible ideas.

Design as language

Laurie Olin has claimed that gardens defy precise satisfactory description or explanation in words.[14] In *The Afterlife of Gardens*, Hunt challenges that assertion.[15] I consider that the real point being made by Olin is that there is a significant degree of apprehension of gardens that cannot be gained if one is reliant on words alone, separated, as they are, from the sensate experience, as a means of analysis. Underpinning the assertion from Olin seems to be that design, and in this case garden design, is itself a means of communication — with its own language. In other words, the means of communication of designers is design, often initially through representation by means of graphics, but ultimately expressed in the organization of the constituent parts of the garden itself. This design language is different in kind from that of words. It lacks the universality of a spoken language shared by a community of speakers. It is a language of a practice and so it is continually evolving, a language based on something that could be described as terms of reference within this field. Though these are argued over within the design community, designers engage in general precepts both as practitioners and theorists. Designers employ ideas, and, as with all languages, their ideas are codified, in this case, in the language of design. They become codified by their need to become an expression in the tangible world. Through the translation from concept to spatial and concrete manifestation in the physical world, ideas are generally unable to hold fidelity to their original intellectual/verbal meaning. Through both the process of design and the process of change from intellectual construct to physical reality, an understanding of concepts may become more generalized or opaque. This design language also tends toward the gestural, has a visual immediacy, communicates complexity and relational characteristics, and operates on many levels and different scales at once. It is not linear or sequential in the way spoken language or traditional scholarly narratives work. Of course, design can be discussed in words (it is done by designers all the time — at least among themselves), but it requires a definite orientation, a differently understood syntax, one that is expressed visually in design form and ultimately in three-dimensions.

The garden itself is a site — an environment and a cultural place. But the design gestures with which it is constructed or endowed, communicate certain characteristics: of relationships, orders, hierarchies, processes, feelings, haptic experiences, even maybe a totality or completeness of its composition, among other things — communicating not just visually, but bodily, triggering emotional responses. This is the *how* of a garden. Gardens *do* things. This doing is the result of design.

Landscapeness

Hunt advocates uniting 'the pen and the pencil (the word and the image) so that words are invoked to enlarge and enhance the full visual and sensual appeal of a garden for its visitors'.[16] Words could be regarded as extra to the *landscapeness* of gardens. I use the term *landscapeness* to describe the materiality and tangibility of gardens — encompassing ground and vegetation and exposure to the elements, and the three-dimensional environment of gardens — being in and of a garden. It could be further argued that, at least as far as the sensual goes, words merely intellectualize something that might otherwise be freely grasped in a sensate manner. Many artists label their work *Untitled* just to avoid any specifics of enhancement or directed bias of interpretation, to permit the work to be appreciated for itself, that is, without verbal prompts.

As in all design or art fields, the landscape of gardens should have the capacity to stand on its own. This *landscapeness* as described above provides the immediacy of the garden experienced. If the immediacy of any garden is a result of its tangibility, anything that may be attributed to a garden by way of cultural or literary references must be as a backdrop. Cultural references are at a remove from the immediacy of sensate qualities as they are usually understood by means of analysis rather than direct experience. Similarly, design intent may have underpinned the original design or informed the design process, but may not necessarily remain evident in the physical outcome of a garden or the evolution of the garden over time. At least in twentieth-century Modernist gardens, other things, such as words and images, could be regarded as addenda to the *landscapeness* of a garden, its tangible and experiential qualities. These additional words and images may, as they often do, detract from the garden itself and its constituent landscape elements and characteristics, or alternatively serve as unnecessary explanations. Hunt states that 'writing something down, especially when the medium is stone, teak, granite or marble, still lends them more than usual importance. If words are "articulate thought", written words are emphatically articulate'.[17] In a context that is essentially non-literary, words — especially emphatically written words — articulate ideas, but may distract from the physical context of the garden and its landscape constituents; in other words, they may distract from its *landscapeness*. Words may lead to an escape from the world of sensate experiences, creating a dislocation from the place and time, the here and now.

Just such a site of words etched in stone and on display is *Little Sparta* in Scotland, with poems and sculpture by Ian Hamilton Finlay (figure 1). These are powerful and evocative poems of social and political discourse, sometimes ironic and humorous. The garden in landscape terms is subservient to the messages inherent in the sculptures. Dominated by the

FIGURE 1. *'The Present Order', sculpture at Little Sparta by Ian Hamilton Finlay. Source: The Estate of Ian Hamilton Finlay. © The Estate of Ian Hamilton Finlay. Reproduced by permission of the Estate of Ian Hamilton Finlay. Permission to reuse must be obtained from the rightsholder.*

sculpture, the landscape is transformed into a sculpture park or an anthology of poems in a landscape. Words, poetry and sculpture are ostensibly objects of social interaction. A garden has its own specifically landscape attributes: tangibly and haptically, it is the materiality of landscape — of ground, plants and water; and ephemerally, the setting outside — of sun, wind, rain, frost, shadows, moonlight, sounds and smells, among many others. These things alone, without the embellishment of references to something else, offer something meaningful. Even avoiding an essentialist approach, these things, such as ground and plants, still have an ability to engage in their own right, and even, as Hunt says, 'enthrall'.[18] They are the materials of a garden designer. Given that garden designers are responsible for such *landscapeness*, they have a creative familiarity with the sensate qualities and experiential immediacy of gardens.

Garden as spatial practice (design)

A garden is a consequence of intent, realized through design. All gardens lie on a continuum between the realization of that intent and its feral otherness, or non-garden. This is constantly in flux. Mara Miller describes gardens as having 'no final form, no unchanging *inert* core'.[19] Nevertheless, despite this ephemerality or fragility, of constant change into a new thing, a garden does have physicality — what Hunt has described as 'an object with "brute-fact" status … being palpably, haptically there'.[20]

A garden's palpable presence in the first instance is based on its unique placement: the *thereness* of any garden cannot be shared by another. It is integral to the nature of the garden as an initial defining point of difference for all gardens. The garden is what it is because of site: to take one element of what constitutes site, the *ground* itself is part of the garden make-up, essential to the nature of the plant life, the flow of water and the traverse of walkers. Secondly, a garden has form. In the language of design, garden design is categorized as a spatial practice, part of a broader field of design practices, and 'form' is usually considered as something encompassed by the term 'spatial design'. This term refers to a variety of characteristics, including: spatial articulation; the relationship between form and space, between materiality and void;[21] edge conditions; ordering principles or organization, whether axes, grids, patterns, symmetry or asymmetry, nodes, threads or organic shapes. Increasingly, more temporal or ephemeral qualities may guide the program: processes, flows, emergence, time spans, intensities, and complexities.[22] Any of these may or may not be in clear evidence in the final design, even if they are an underpinning ordering design principle. Laurie Olin writes that 'everything that exists has form'.[23] So, despite gardens becoming something new all the time, at any one moment they have form.

El Novillero: a case study

Returning to Barnett's claim that gardens of the Modernist era had meaning, this paper addresses one of the seminal works of this era of garden building to discuss how meaning is derived from the material of landscape and its design rather than external cultural references. El Novillero, in Sonoma County, California, north of San Francisco, was designed by Thomas Church in 1949 for the Donnell family and which usually goes by the Donnell name. I have chosen this garden because it epitomizes those gardens of the mid–twentieth century often considered to be solely about form and, as a consequence, to be without meaning — those reappraised by Barnett. Of course, this garden can be placed in developmental terms chronologically with Modernist gardens in France, the UK, and California, and also in relation to social and lifestyle characteristics of the West Coast of the USA. In addition, this garden can provide us with an example of what spatial design is, how it works, and what role its physicality and spatial design may play in a more comprehensive understanding of it.

The area of interest usually discussed in the literature on this garden is the pool (figure 2). The prime generator of the design of this garden could be considered the Modernist idea of functional space — in this case, a response to the hot, dry Californian environment and the partying social climate. It is based on outdoor living, focused around a swimming pool, with lanai, or pool house, a diving board, changing sheds and expansive paving, accoutrements enabling the Californian lifestyle of swimming, socializing, conversing,

FIGURE 2.   *The pool area as an outdoor living and play space. Source: photograph by author.*

sunbathing and drying. So, in its time, these collective components signalled outdoor activity and a play space, probably like no other before it. Their arrangement, layout, and the spatial organization of this area, reinforce the functional aspect of the design. On these points, much has been already written, primarily by a small group of designers and historians, who have focused on Modernist landscapes.[24]

New findings

The following additional points of interpretation are based on a visit to the garden on 23 June 2010. I approached the garden with some understandings already, thanks to prior reading about the garden, and developed further understanding based on my design background. This section constitutes a series of interpretations based on this experience, focused on the 'formal' characteristics (as framed above), with an eye to the design intent as evidenced on the ground. These are possible interpretations intended to contribute to an ongoing range of interpretations.

My sensate experience of the garden included a long drive through dry pastureland before arrival at what seemed a green oasis. The glade of live oaks is shady and cool, and provides some enclosure before arrival at the pool terrace, open and sunny with views outward, the pool an emerald colour. The description of this experience confirms Olin's point that words cannot fully capture a full or precise understanding. In addition, this experience was only mine and only on that day.[25] However, from a design perspective, I am able to discuss and analyse the materiality, structure and apparent design intent.

Further to the above description, the layout of the pool garden includes the background of a stone wall forming an edge to this space, partially enclosing it. This stone wall is continuous between garden edge and lanai, forming the back wall of this building. The plan drawing (figure 3) shows the wall as the heavy line wrapping along the west side of the patio pool area and the almost invisible line of the glass walls of the lanai. Though there were some minor peripheral differences between what was built and this earlier version of the design (as above), the design intent is clear. Figure 4 demonstrates the continuity of the wall as both garden-defining edge and lanai wall, with no apparent difference in its treatment (height, materiality, un-modified line)

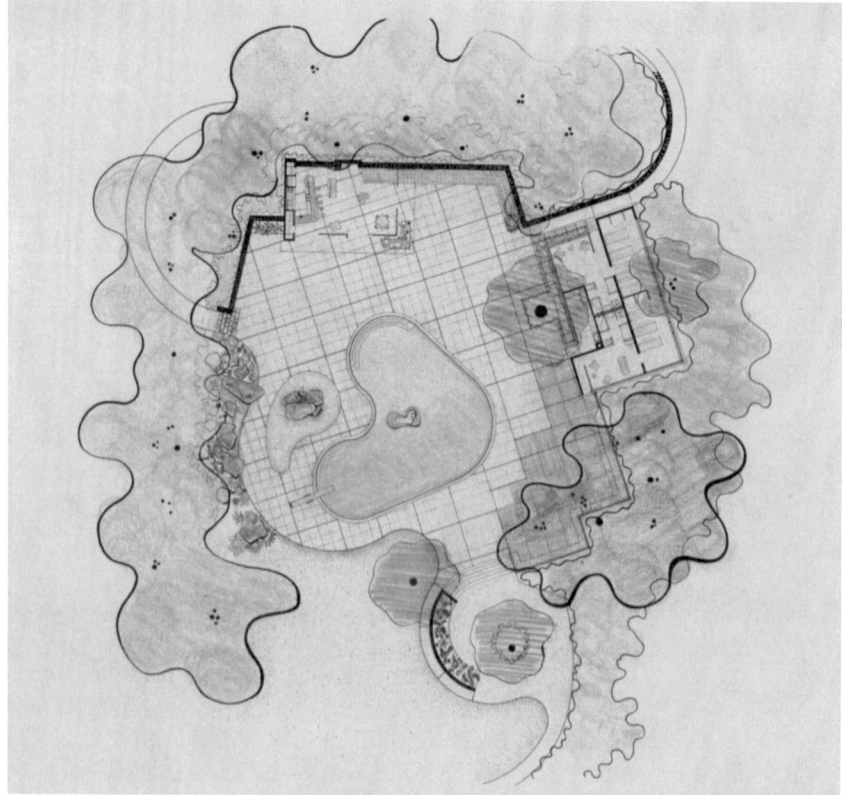

FIGURE 3. *Dewey Donnell residence pool area design. Source: Environmental Design Archives, University of California, Berkeley. © Regents of the University of California. Reproduced by permission of the Thomas D. Church Collection, Environmental Design Archives, University of California, Berkeley. Permission to reuse must be obtained from the rightsholder.*

between the two. This indicates, too, that the primary function of this wall is as an edge to the whole garden area, inclusive of building and open space.

The rest of the lanai has glass walls, transparent or completely open from slab roof to floor. The floor is continuous between outdoor and indoor in material and levels. So, this room, the most important for the outdoor lifestyle centred on the pool, is fully integrated with the garden, indeed, is almost invisible as a room, but for its overhead plane. In a sense, it has become part of the garden.

FIGURE 4. *Rear wall and lanai. Source: photograph by author.*

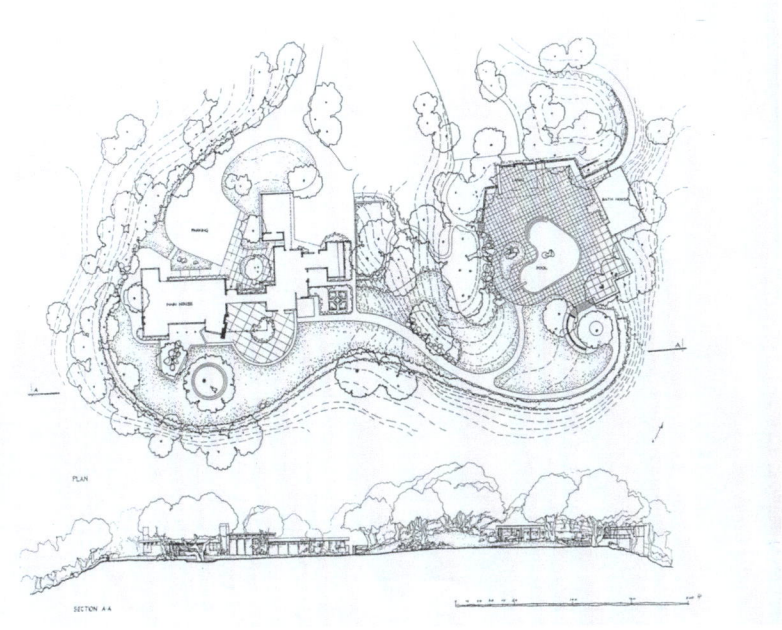

FIGURE 5. *Dewey Donnell residence site plan and elevation. Source: Environmental Design Archives, University of California, Berkeley. © Regents of the University of California. Reproduced by permission of the Thomas D. Church Collection, Environmental Design Archives, University of California, Berkeley. Permission to reuse must be obtained from the rightsholder.*

Taking into consideration the layout of the whole site, as well as the way in which it developed chronologically with the house, this function demonstrates an even more definitive spatial configuration. Although the house and pool area are not separated by a great distance, they are by topography, vegetation, and lack of sightlines (see figure 5, plan and section). Treib describes the site as a garden in two parts.[26] In addition and importantly, the garden was built before the house by a number of years. During these years, prior to the main house being built, the lanai was used as a living room, guests were put up in the changing rooms, and the pool garden was used for parties. Interestingly, one of the rooms in the house built later is full of plants, has glass walls, 'wettable' surface and outdoor-style furniture. In other words, a garden has been sited within the house, and the pool garden has become a living space in a house-like manner. More than simply outdoor living, the pool area acts operationally and resembles structurally the equivalent of a dwelling. Circumstance and design intent have created a new type of space.

As mentioned already, this garden and others like it were considered to be all about form. This garden, as with others, has gained recognition as having utilized the forms common to Modernist art and associated design objects. Treib: 'by 1947, the biomorphic shapes of painters such as Joan Miró and sculptors such as Isamu Noguchi were widely known, and kidneys, amoebas and boomerangs were all part and parcel of the motifs of postwar America'.[27] Treib also speculates on possibilities of various other influences acting on Church and the sources of inspiration for this particular garden, including the architecture of Alvar Alto. In his earlier work, 'Axioms for a Modern Landscape Architecture',[28] Treib discusses at some length the artistic influences bearing upon Church, including Cubism and Surrealism, concluding that they may have contributed more to an undoing of prior forms than the establishment of a new vocabulary of forms. Many others have declared that by copying the forms of an external practice — art in the form of Cubism and Surrealism — garden design, and even the

FIGURE 6. *The landscape context of dry grasslands, evergreen oaks,* Quercus agrifolia, *rocks, serpentine sloughs of the Sonoma valley river system. Source: photograph by author.*

broader field of landscape architecture, was denied an interpretation based on its own landscape condition.[29] Steven Krog writes: '[w]ith the best of intentions, landscape architecture has appropriated the images of modern art and oriental gardens but — out of ignorance, convenience, or deliberation — failed to comprehend the ideas that generated those images'.[30]

What may be much more pertinent is the use to which Church put these forms. If we look at the site conditions Thomas Church faced, we see dry grasslands, evergreen oaks (live oaks), *Quercus agrifolia*, rocks, and the serpentine sloughs of the Sonoma valley river system (figure 6). The apparent reiteration of some Jean Arp or Miró-style artwork seems a somewhat formalist object in plan form, as seen in figure 2, even if it has a reasonably dynamic relationship to its neighbouring forms. However, when viewed from the position of sitting in or just outside the lanai, the line of the pool's edge appears to echo the forms of the waterways winding across the valley that this terrace overlooks (figure 7). This point, too, has already been

FIGURE 7. *The well-known shape of the pool as seen from the lanai. Source: photograph by author.*

FIGURE 8. *Framed live oaks. Source: photograph by author.*

FIGURE 9. *Local rocks. Source: photograph by author.*

discussed in the literature.[31] The very site responsiveness of this form makes all other kidney-shaped pools, which copied this one, almost certainly incongruous with their setting, unless they too have serpentine waterways against which they can be sighted. The patio pavers are the sandy colour of the dry California grasslands, and the rocks, artfully clumped together, are sited on grass forms like sculptures on their pedestals, as are the live oaks in their timber frames (figures 8 and 9).

El Novillero's significance

It is the site conditions which Church has grounded in this design: together, pool-edge shape, pavers, rocks and trees, establish a comprehensive set of designed conditions which demonstrate a deliberate abstraction of the site's raw material. It may be that at least some of the forms of this garden were derived from those of early twentieth-century art, as Treib and others contend. But, more importantly, they reflect the specificity of the local. Whether they are forms derived from the fashionable motifs of the time, or Cubist or Surrealist art gestures, the significance lies in their abstracted representations of existing landscape conditions. The abstraction reveals much more than form copying: the design signals content and context over form, but form, which allowed Church to express the characteristics of this place in an abstract way.

Representing the local as a stylized landscape in the abstract — motifs expressed in pure forms — allows this garden to connote all gardens. Its materiality of ground, vegetation and rocks, and sensate qualities are shared by all gardens, and its localness is both unique and a fact inherent to all gardens. Barnett claims that since the nineteenth century:

> [T]he landscape architect whose body of knowledge ranged across both the arts and the sciences felt ideally situated to develop a comprehensive synthesis of the universal and the particular on that very threshold 'where culture meets nature' — in the private garden.[32]

Paradoxically, abstraction of the particular of this garden's site features makes apparent the universal. The pool here might be functionally a play space and its sweeping edge might be a simulacrum of the river channels below, but its stylized line may also capture in a single gesture the manner in which water behaves. Likewise, though the rocks, trees and dry grasslands are present on this site as dominant features, they are also essential landscape characteristics representing fundamental elements of the idea of landscape. Barnett writes that, like artists of the time, Modernist landscape architects 'struggle[d] with representation and with the questions of truth and the relation of the eternal to the fleeting moment'.[33] He asserts that abstraction was a concern of Modernist art in pursuit of the Absolute in a secular society. El Novillero appears to be a tangible landscape example of this endeavour. Without reference to outside phenomena, except an incidental borrowing of a recognizable contemporary form, this garden expresses eternal *landscapeness* based on what is present. In addition, little or no explanation is offered by the designer of any meaning or intangible ideas (noumena in Hunt's terms). Only in this unspoken, but eloquent, manner is this garden a signifier of anything other than itself.

A postscript is required to this garden composition. In terms of the *longue durée*, the native live oaks, *Quercus agrifolia*, are gradually dying, probably as a result of watering to maintain the verdant green of the lawn. As they die, they are being replaced by *Quercus virginiana*, a native of the eastern seaboard, tolerant of these wetter conditions.[34] The implications are that the materials, which were indigenous, and the design which echoed the local, have been compromised, and the very indigeneity at the core of the original composition will be lost in time.

Conclusion

The study of gardens needs first to take into account the basic characteristics of gardens themselves, their physicality, the experience of them by visitors and their design intent. After all, one of the most potent of *raisons d'être* for garden creation is for visitors to be enveloped, even enthralled, by the sensate qualities of the garden itself. These sensate qualities are at the core of the idea of garden. It may be, as Olin claims, that the experience of the sensate qualities of a garden are beyond satisfactory explanation, but their effects surely are not beyond discussion. Often histories of gardens have avoided commentary on these aspects of gardens, and nowhere is this more apparent than in accounts of mid-twentieth-century, so-called Modernist, gardens. While practitioners, like Church, did not expressly place meaning to their forms, sound analysis from a design perspective can establish comprehensive understanding of their gardens. Site conditions, intent (even if speculative) and its manifestation into the material of landscape, potential visitor experience, spatial articulation and relationships are the material and tools of landscape designers. Such multiple layers of input allow many alternative interpretations. Other garden designers could offer a different analysis of gardens, particularly for those that may be considered of some importance, from the design perspectives of form, spatial design, materiality, hapticity and experience. It may be that Modernist gardens that were inclined to utilize the forms of other fields of art and design, and whose designers were reluctant to discuss meaning, are more likely to yield new and worthwhile interpretations, revealing a range of intrinsic meanings. But all gardens should be deserving of this sort of analysis. If the historiography of gardens can include analysis from a design perspective, it can become a richer field, and signification discovered where none was previously recognized.

Disclosure statement

No potential conflict of interest was reported by the author.

NOTES

1. Rod Barnett, 'Gardens Without Meaning', *Landscape Review*, iii/2, 1997, pp. 22–42.
2. Although Barnett acknowledges the difficulty of defining Modernism, for convenience he situates the gardens with this designation 'in the fast, disorienting decades that encompassed the world wars of this century', p. 23.
3. Barnett, 'Gardens Without Meaning'.
4. For a relatively recent collection of contributors to this ongoing discussion, see Marc Treib (ed.), *Meaning in Landscape Architecture and Gardens: Four Essays, Four Commentaries* (Abingdon, UK: Routledge, 2011). This includes four collated essays originally published between 1988 and 2007, and forming the basis of a Council of Educators in Landscape Architecture forum in 2009. See also, the book's review by Jacky Bowring, 'Meaning in Landscape Architecture and Gardens', *Landscape Review*, xiii/2, 2011, pp. 40–43.
5. Rebecca Solnick, 'The Colour of Shadows, the Weight of Breath, the Sound of Dust', in *Once Removed: Portraits by J. John Priola* (Santa Fe: Arena Editions, 1998), pp. 113–123.
6. Clarence Glacken, *Traces on the Rhodian Shore* (London: Oxford University Press, 1967), p. 348.
7. John Dixon Hunt, *Greater Perfections: The Practice of Garden Theory* (London: Thames & Hudson, 2000).
8. Barnett, 'Gardens Without Meaning', p. 24.
9. Stanislaus Fung, 'Longing and Belonging in Chinese Garden History', in Michel Conan (ed.), *Perspectives on Garden Histories* (Washington, DC: Dumbarton Oaks Research Library and Collection, 1999), pp. 205–219; Xu Yingong, 'Gardens as Cultural Memory in Suzhou, Eleventh to Nineteenth Centuries', in Michel Conan and Chen Wanheng (eds), *Gardens, City Life and Culture* (Washington, DC: Dumbarton Oaks Research Library and Collection, 2008), pp. 203–227.
10. For a comprehensive scholarly treatise on Japanese *karesansui* (dry raked gardens), see Wybe Kuitert, *Themes in the History of Japanese Garden Art* (Honolulu: University of Hawai'i Press, 2002).
11. John Dixon Hunt, 'Approaches (New and Old) to Garden History', in Conan (ed.), *Perspectives on Garden Histories*, p. 90.
12. John Dixon Hunt, 'Verbal Versus Visual Meanings in Garden History: The Case of Rousham', in John Dixon Hunt (ed.), *Garden History: Issues, Approaches, Methods* (Washington, DC: Dumbarton Oaks Research Library and Collection, 1992), p. 152.
13. Marc Treib, 'Sources of Significance: The Garden in Our Time' in Stuart Wrede and William Howard Adams (eds), *Denatured Visions: Landscape and Culture in the Twentieth Century* (New York: Museum of Modern Art, 1991), pp. 106–109.
14. Laurie Olin, 'What Did I Mean Then or Now?' in Treib (ed.), *Meaning in Landscape Architecture and Gardens*, pp. 72–81.
15. John Dixon Hunt, *The Afterlife of Gardens* (London: Reaktion Books, 2004).
16. Hunt, *Afterlife of Gardens*, p. 92.
17. Hunt, *Afterlife of Gardens*, p. 102. The phrase 'articulate thought' is a quote from Indra Kagis McEwen, *Vitruvius: Writing the Body of Architecture* (Cambridge, MA: MIT Press, 2002).
18. Hunt, *Afterlife of Gardens*, p. 48. Note also Hunt, *Nature Over Again: The Garden Art of Ian Hamilton Finlay* (London: Reaktion Books, 2008), in which he describes the spatial and tangible qualities of this garden along with the comprehensive role played by words.
19. Mara Miller, 'Time and Temporality in Japanese Gardens', in Jan Birksted (ed.), *Relating Architecture to Landscape* (New York: Routledge, 1999), pp. 43–58.
20. Hunt, *Afterlife of Gardens*, pp. 15–16.
21. See a comprehensive explanation of landscape's materiality and permanent presence and corresponding lack of *tabula rasa* in Peter Connolly, 'What Is at Hand? A Re-evaluation of Technique in Landscape Architecture', *Kerb: Journal of Landscape Architecture*, vi, 1999, pp. 70–83.
22. Among a significant number of authors who place importance on landscape processes are: Hunt, *Greater Perfections*, p. 9; Julian Raxworthy, 'Landscape Symphonies: Gardening as a Source of Landscape Architectural Practice, Engaged with Change', paper presented at Society of Architectural Historians, Australia and New Zealand (Melbourne, Australia, 2003); Christopher Griffiths, *Reframing the Given: Landscape Analysis Techniques* (Saarbrücken: VDM Verlag, 2008).
23. Laurie Olin, 'Form, Meaning and Expression', in Simon Swaffield (ed.), *Theory in Landscape Architecture: A Reader* (Philadelphia, PA: University of Pennsylvania Press, 2002), pp. 77–80.
24. Notably, Marc Treib, 'Axioms for a Modern Landscape Architecture', in Marc Treib (ed.), *Modern Landscape Architecture: A Critical Review* (Cambridge, MA: MIT Press, 1989), pp. 36–67; the following issue: *Studies in the History of Gardens and Designed Landscapes*, xx/2, 2000; Jory Johnson and Felice Frankel, *Modern Landscape Architecture: Redefining the Garden* (New York: Abberville Press, 1991).

25. Hunt in *The Afterlife of Gardens* discusses the contribution of many and different visitations to gardens.

26. Marc Treib, 'Thomas Church: The Modernist Years', *Studies in the History of Gardens and Designed Landscapes*, xx/2, 2000, p. 140.

27. Treib, 'Thomas Church: The Modernist Years', p. 145.

28. Treib, 'Axioms for a Modern Landscape Architecture', pp. 36–67.

29. Barnett, 'Gardens Without Meaning', p. 27, includes commentary to this effect from John Dixon Hunt, *Gardens and the Picturesque* (Cambridge, MA: MIT Press, 1992).

30. Steven Krog, 'Wither the Garden?', in Wrede and Adams (eds), *Denatured Visions*, p. 96.

31. See Treib, 'Axioms for a Modern Landscape Architecture', pp. 36–67, and Treib, 'Thomas Church: The Modernist Years', pp. 130–156, in both of which he treats this interpretation as hearsay; see also, Marc Treib, *The Donnell and Eckbo Gardens: Modern Californian Masterworks* (San Francisco, CA: William Stout Publishers, 2005), in which he quotes Church regarding the inspiration from the waterways; see also, Peter Walker and Melanie Simo, *Invisible Gardens: The Search for Modernism in the American Landscape* (Cambridge, MA: MIT Press, 1994), p. 110.

32. Barnett, 'Gardens Without Meaning', p. 31.

33. Barnett, 'Gardens Without Meaning', p. 32.

34. J. Faggiolo (member of the Donnell family), 2010, personal communication.

Rethinking Australian natural gardens and national identity, 1950–1979

CHRISTINA DYSON

In Australia following the Second World War, shifting dynamics between Britain and Australia alongside rapid social transformation and large-scale environmental change brought significant changes to Australian cultural life, prompting the renegotiation of Australia's national identity. Garden historians are increasingly aware that during this same period a new kind of garden also began to emerge in Australia.[1] Over an approximately three-decade period spanning the 1950s through the 1970s, garden writers and garden and landscape designers began exploring new concepts for Australian gardens and designed landscapes. Seeking a design ethos that would be recognized as identifiably Australian, they looked to Australian native flora and indigenous landscape themes for inspiration. The gardens that emerged differed markedly from the gardens and gardening practices that were part of Australia's Anglo-European cultural inheritance. They have been described anecdotally as 'backyards assiduously cultivated to mimic the Australian bush'.[2] They formed a distinctive sequence of quadrilateral shapes amongst a complex suburban patchwork of 'micro-ecosystems' which mirrored the diverse heritages and traditions of Australia's post-war Anglo-Australian and migrant population.[3]

Unifying this diversity of gardening patterns in the post-war suburban landscape is the function of the garden as a space in which associations with specific *homelands*, cultures, and communities could be imagined and constructed.[4] This follows historian Anne Helmreich's conception of the garden as both cultural product and 'artefact' of a particular set of cultural practices.[5] The imagined gardens and the gardening literature which form the focus of this article also functioned as spaces in which a relationship with the natural Australian environment could be explored and developed, and in which differentiation from a remote British culture could be articulated. In this regard, they functioned as a kind of 'contact zone', in the sense defined by Mary Louise Pratt 'as social spaces where cultures' come together, whereby the garden provides a space in which relationships between cultures and between cultures and place can be negotiated.[6]

National identity theory asserts that formulating understandings of national identity must contend with both geographical *and* historical context.[7] Attention to the historical dimensions of national identity formulation in the context of the post-war Australian natural garden represents an under-researched area of Australia's garden history.[8] Historians, to date, have instead foregrounded the function served by the distinctive material and experiential attributes of Australian flora and the indigenous landscape.[9] In extending that scholarship, I contend here that there was a further — and unacknowledged — role played by Australian flora and landscape during the 1950s to 1970s. I argue that in addition to material distinctiveness, as others have identified, the deep sense of time that Australian flora and landscape were imagined and scientifically asserted to embody was central to their role in the renegotiation of national identity and to the articulation of a distinctly national culture. The scope of this article is therefore intentionally narrow and it will draw on national identity theory in order to uncover the significance of the function that history served in this particular time and space.[10]

Increased Australian plant-content in the natural gardens espoused and produced through the period from 1950 to 1979 made them distinctive relative to their historical and suburban contexts, but not necessarily unprecedented.[11] Australian native plants had been utilized in Australian gardens and horticulture from the colonial period.[12] While promoted and utilized for different reasons, some key attributes shared among garden-makers from the colonial to interwar periods include identification with, and attachment to, Britain and the 'old world' as homeland, and the notion that Australian plants conformed to other pre-established garden or parkland forms and aesthetics.[13]

From colonial times and into the 1930s, European Australians tended to rely upon inherited British traditions for a sense of a common heritage, cultural background, and as benchmarks for cultural sophistication — indeed, to some extent, this remains the case. Increasing substitution of Australian plants in place of non-Australian species into conventional and inherited gardening patterns into the interwar period continued this tendency. By the late 1930s, desires for cultural maturity independent from Britain emerged in Australian art and architectural circles.[14] Yet there was a lag in the development of such desires in Australian architecture and landscape architecture.[15] Not until the mid-1950s and 1960s did similar desires begin to be articulated in garden and landscape design writing, by which time the physical remnants of Australia's vernacular rural past — which had, from the late nineteenth century, served notions of a national heritage[16] — sat uncomfortably with the ambitions of landscape designers for cultural sophistication. From the 1960s, enthusiasm for Australian native plant gardens similarly lacked prestige, still less the desired cultural maturity associated as they were with parochialism, idealistic purism, and nationalism.[17] Instead, in most cases, it was 'the bush' of awakened environmental sensibilities and its manipulation that were sought in attempts to define a sophisticated and distinctly Australian design vocabulary. Remnants of older imagery and symbolism, however, seeped into these new ideas, which says much about the power of myth and its tendency to be translated into new eras and new contexts.

If the use of Australian plants dated back to colonial times, their post-war presence in gardens differed markedly from what went before because of corresponding changes in design concepts. Holmes describes the post-war use of Australian plants in gardens as 'a radical rethinking of the garden aesthetic then dominant in Australia'.[18] Such a rethinking encouraged close observation of the aesthetic qualities and planting structures of the natural environment, as opposed to merely substituting Australian plant species for introduced varieties within conventional, inherited gardening traditions and patterns, as had earlier occurred.[19]

Manipulating history in Australia's post-war natural garden

Landscape and garden designers promoting natural garden design in post-war Australia manipulated history to legitimize their new conceptions. One of the first of those was in 1952, when prominent garden writer, garden designer, conservationist and photographer Edna Walling (1895–1973) published *The Australian Roadside*; popularly identified as the moment at which Walling shifted her interest from cottage to native plant gardens.[20] *The Australian Roadside* contains numerous photographs by Walling of romanticized pastoral landscapes and meandering, tree-lined, rural country lanes presented as places where time stood still (figure 1). For their nostalgia, sentimentality, and value placed on 'tangible reminders of European Australia's beginnings' as a national heritage, Walling's imagery shared much with the late-nineteenth-century antiquarians and the 'nationalist aesthetics' of early twentieth-century nature writers.[21] In order to construct a national heritage and culture, both were mining Australia's recent European history embodied in 'the Bush' — meaning pastoral landscapes strewn with 'noble gum trees, [and] peopled by idealised shearers and drovers'.[22] This construct of 'the Bush' as the dominant and symbolic image of Australia had its origins in the late nineteenth century.[23] Yet, as shown by Walling's *The Australian Roadside*, and even more recently by the opening ceremony at the Sydney 2000 Olympic Games, it has remained a powerful and popular nation-defining image well into the twentieth and twenty-first centuries.[24]

Like Australia's nature writing of the 1920s and 1930s,[25] Walling's *The Australian Roadside* of 1952 was also a published argument for the preservation of aspects of Australia's national heritage embodied in both rural scenes and primitive areas, more than it was a book on garden design. It captured the spirit of increasing environmental awareness of the time, which saw the formation of local community preservation groups from the early 1950s like the Beaumaris Tree Preservation Society (estab. 1953) in suburban Melbourne, whose promotional literature cited Walling's *The Australian Roadside* extensively,[26] and state and local-level provisions for nature reserves and wildflower sanctuaries.[27]

FIGURE 1. *Walling's caption to her photograph (adhered to verso) reads: 'The wise serenity of strong Red Gums: The gate made simply for generations to use, the coolness of flower-sprinkled grass, breathe an atmosphere of deep-rooted peace, where the Earth and her fruits, the seasons and their needs and gifts, alone mark passing time'. Source: State Library of Victoria. © The Estate of Edna Walling.*

The Australian Roadside also reveals Walling's belief in publically visible landscapes as crucial sites for expressing the identity of a nation and its citizens.[28] Walling extended this ideology to the designed landscape three years later in 1955 in *The Age*, one of Melbourne's major daily newspapers.[29] In relation to the development of a large-scale civic landscape-design project associated with the 1956 Melbourne Olympic Games — an international event anticipated as making Melbourne and the nation visible to the eyes of the world — Walling agitated for Australian tree species and their arrangement in groupings that would 'produce the effect of natural landscape planting' and

replicate 'the primitive beauty of the Australian landscape'. In drawing on the informal planting structures of natural landscapes Walling was promoting a narrower kind of bush than she had depicted in *The Australian Roadside*, one, in this case, as an appropriate idiom for urban landscape design. Libby Robin has distinguished between these two conceptions of 'the Bush' as 'a pastoral frontier' and 'the bush' as 'the wilderness of environmentalism'.[30] The latter can be connected to increasing environmental awareness from the 1950s and, most prominently, to emerging ecological consciousness around the mid-1960s.[31] Conceived as ancient and pre-dating European colonization, Walling now reached further back to a primitive natural landscape conceptualized as free from the degeneracy of human destructiveness,[32] epitomized by the bulldozer and so-called 'progress'.[33] This again reflected Walling's conservationist sensibility, but the primitive also bestowed the sanction of precedent on her design concept, itself a radical departure from the traditional character of Melbourne's other parks and gardens. Since the nineteenth century, Melbourne's human-constructed, urban green spaces had played a central role in fashioning an image of the city as civilized, cultured, and European.[34]

A sense of the past was also important in the gardens of landscape designer Ellis Stones (1895–1975). Stones began his landscape-design career working for Walling in Melbourne in 1932.[35] He also built his own gardens and ultimately developed his own landscape-design practice which he continued until his death in 1975. Stones increasingly used Australian plants in his garden designs from the 1930s (though not exclusively)[36] and his gardens were admired for their Australian native landscape 'feel'. A contemporary of Stones', Alistair Knox, reflected in 1980 that ' ... Ellis could visualize the worn-down hills, the ancient boulders and the "morning of the world" plant life he remembered from his boyhood, and could then put his thoughts into action'.[37] For Stones, the weathered character of boulders in their natural state was important in his gardens, and is demonstrated in his designs for the University of Melbourne at its Burnley campus (the Ellis Stones Rockery, 1962) and Parkville campus (the native garden in the south-west corner of the South Lawn Precinct, 1972–1973). The bold use of stone in these gardens, and others by Stones, suggest the intent for them to embody a sense of stability and belonging; a sense of having always been there and a sense that the garden and its owners or custodians were there to stay (figure 2). In *Australian Garden Design* (1971), Stones wrote:

I always consider weathered boulders one of the most valuable materials for use in the garden, or the natural landscape ... The stone selected [for artificial rock work] ... should be surface stone that has weathered, with perhaps lichen and moss growing on it.[38]

This belief suggests that Stones also valued continuity with the past. Weathered boulders provided tangible reminders of, and links to, the rural landscapes of his boyhood Euroa, in northeast Victoria. But they also signified much older landscapes. Through close observation of natural processes and accretions in the landscape, Stones embraced an anti-escapist notion of the past which distinguishes his manipulation of the past from Walling's. Unlike Walling's romanticized pastoral landscapes as places where time stood still, or primitive landscapes of the distant past, Stones embraced a 'living past bound up with the present'.[39] Lichen, moss, and evidence of weathering were important as tangible signs of age. These accretions embodied the passage of time, providing a sense of continuity between past and present whereby the sense of accretion enriched both the past and the present (see also Jacky Bowring's article in this special issue).[40]

Although he lived in Ivanhoe in suburban Melbourne, Stones developed strong connections with members of an alternative community of artists, writers, and other creative people in Eltham, some 15 km away, including landscape designers who shared an environmental sensitivity and sensibility. Environmental architect Alistair Knox (1912–1986) was an integral member of the Eltham community[41] and a leading figure in environmental design in Melbourne. Knox championed and developed architecture and landscape design in harmony with the environment during the period considered in this account.[42] The 'environment' that Knox valued was a palimpsest of remnant indigenous landscapes and tangible reminders of the area's European settlement history — clay roads, 100-year-old exotic trees, the rolling hills of pastoral landscapes. It was a view that reflected the breadth of his conservation concerns.[43]

In a paper delivered at the first conference of the newly formed Australian Institute of Landscape Architects in 1969 in Sydney, Knox presented the merits of the indigenous landscape as a concept to applied landscape design.[44] In this paper, Knox conjured romanticized scenes of an ancient and timeless natural landscape. He referred to ' ... the gang gangs flopping through the primeval and timeless bush ... ', ' ... the ageless river red gum', and 'the antedeluvian

FIGURE 2. *The rockery designed by Ellis Stones in 1962 (restored 2002–2003) at Burnley Gardens, The University of Melbourne, demonstrates his fondness for weathered stone in the gardens he designed and made. Source: photograph by author (2009).*

[sic] iron bark … '.[45] In later publications by Knox, references to antiquity and the primeval were intermingled with different and sometimes contradictory conceptions of the scale of Australia's environmental and human past. These conceptions included European Australia's recent history, in *Living in the Environment* (1975),[46] and personal memory, in *We Are What We Stand On* (1980). Through recourse to 'the Bush' associated with European Australia's mythologized pioneering ancestors, Knox drew on a version of the pioneering settlers and bushmen from Australian history, described by John Hirst in his essay 'The Pioneer Legend' (1978).[47] Through these figures, Knox imagined the harshness of the Australian environment as ' … the great constructor and unifier of the Australian character'.[48] Subscribing to the common belief that nations in modern times are forged in conflict — 'man against man, country against country, ideology against ideology' — for Knox, the Australian character and the nation were formed through a different kind of conflict; 'man against the elements'.[49] While David Lowenthal has argued that 'valued traits … attributed to the past are seldom consciously identified',[50] Knox explicitly linked aspects of the past he conceived and valued in landscape to identity making.[51] The intermingling of different meanings of the bush in Knox's conceptions of the past in that landscape suggest both were valued, and that perhaps the surviving elements from European Australia's settlement past had become naturalized into the indigenous landscape. Of central importance to Knox's environmental aesthetic was its differentiation from suburban uniformity, and the perceived absence of meaning in suburbia.

Pursuing similar goals to Knox in creating gardens and landscape designs that would be in harmony with the environment, yet independently from the activity in Melbourne with which Knox and Stones were associated, were Sydney-based artists and garden designers Jean Walker (b. 1922) and Betty Maloney (1925–2001). Maloney and Walker's first successful published argument for 'design in harmony with our own very wonderful environment' appeared in *Designing Australian Bush Gardens* (1966).[52] A contemporary reviewer described the impact of this work on gardening patterns as 'having aroused a furore of interest … so welcome … [and] at once an inspiration and a bomb'.[53] The bush in which Walker and Maloney's design concept found its inspiration was the Hawkesbury sandstone ecology that was under threat within the bushland suburbs of northern Sydney. This was not 'the Bush' of the pastoral frontier. It was closer to the 'wilderness of environmentalism'; although in truth it was not 'wilderness' at all, but natural and seemingly natural bushland in urban and suburban contexts.[54]

Published in 1967, *More About Bush Gardens*, also by Maloney and Walker, was unequivocal in its awareness of, and wonder at, the primitive age of Australian landscape and its living embodiment in the local flora.[55] *Designing Australian Bush Gardens* also celebrated the 'timeless tranquillity' of the Australian bush and the 'very ancient forms' of particular plants. In *More About Bush Gardens*, however, antiquity and continuity with the past form major themes. These themes are present throughout in multiple separate references to plants as primitive, archaic, and unique within an ancient and remote land. Several plant types, among them ferns and banksias (figures 3 and 4), are also imagined as embodying continuity with the ancient past, described as 'living fossils' and sketched as gnarled and worn by time and by the elements.[56]

As well as Australian native plants, well-weathered stone and boulders were important elements in Maloney and Walker's bush gardens. These functioned as tangible embodiments of much older landscapes. Like the value placed on weathered boulders by Ellis Stones, for Maloney and Walker, stone and boulders symbolized stability and intransience, and its use was encouraged for the feelings of 'strength, permanence and grandeur' it brought to the bush garden. Stone was also valued as a symbol of the antiquity of 'Australia [as] a very old place'. '[T]ime stood still for our continent … ', wrote Maloney and Walker, who conceptualized these as the geological equivalents of the grass tree: ' … both are primitive, both ancient'.[57]

The introductory words to *More About Bush Gardens* reveal the authors' awareness of scientific speculation about the age and remoteness of the Australian flora and, for its lineage with that past, as unique on a world scale:

> Australia's garden was born before the dreamtime, sea-framed and left to drift in loneliness, as pendulous guardian of the primeval. The dwarf pine of Tasmania … the archaic fork-ferns, simple, flowerless, fern-like plants which bridged the waters to dry land and dominated the world's first forests, are still here though long since turned to stone in other parts of the world.[58]

Recognition of the ancient age of the Australian landscape and flora was not new to post-war Australia. Scientists such as Joseph Hooker in his *Flora Tasmaniæ* (1860) and Professor A. C. Seward in his paper presented to the

FIGURE 3. *Fern garden (plan) incorporating large boulders by Jean Walker. Source: Betty Maloney and Jean Walker,* More About Bush Gardens *(Sydney: Horwitz Publications Inc. Pty Ltd, 1967), pp. 16–17. © Jean Walker. Reproduced by permission of Elizabeth T. Smith and Ian and Rohan Walker, on behalf of Jean Walker. Permission to reuse must be obtained from the rightsholder.*

1914 Australian meeting of the British Association for the Advancement of Science, 'The Vegetation of Gondwana Land', had earlier suggested Australia's flora was a relic of a different era or a diverse range of connections;[59] and their ideas continued to be debated well into the twentieth century.[60] Accessible information about the age of the Australian continent and its flora was also circulating by the 1930s in journals like *The Victorian Naturalist*[61] and, by the late 1940s and early 1950s, in popular natural history magazines such as *Wild Life* (later *Wild Life and Outdoors*).[62] In 1938, Thistle Harris published *Wild Flowers of Australia*, a popular guidebook intended to provide accessible information about Australian flora to a wide audience.[63] The introduction to *Wild Flowers of Australia* highlighted Australia's geological past and introduced the

notion that Australia may once have been 'connected with northern lands through New Guinea', and some of the 'brush' plants of eastern Australia as 'relics of the first flowering plants ... '.[64] Although *Wild Flowers* is attributed to Harris, the original manuscript was prepared by Edwin Cheel of the Sydney Botanic Gardens. Cheel attended the 1914 Australian meeting of the British Association for the Advancement of Science.[65] Harris's *Wild Flowers of Australia* apparently impressed Edna Walling when it was first published,[66] and was also an important plant reference for Maloney and Walker.[67]

Publications by nature writers like Elyne Mitchell in *Australian Treescapes* and botanist James W. C. Audas (1872–1959) in *The Australian Bushland*, both published in 1950 for general readerships, also helped to popularize the notion of Australia as an ancient land and the idea that particular plant species — especially, aged and statuesque eucalypts like the river red gum and the mountain ash, dryandra, waratah, for example — provided continuity between the past and the present.[68] In her essay of the same title in *Australian Treescapes*, Mitchell pinned 'the quintessence of Australia' to ' ... the immense space of the ancient land that still bears, in all its trees, and plants, and animals, in the earth itself, the imprint of creation and eternity'.[69] Essays such as Mitchell's were important for providing a bridge between scientific knowledge of the Australia's natural environment and more general readerships. The idea of 'fossil plants' was further promulgated in the 1956 nursery catalogue of one of the earliest specialist native plant growers, George Althofer, who provided native plants by mail order throughout Australia.[70]

Ancientness and continuity with the past are separate attributes valued in the past.[71] Maloney and Walker valued both. Like Walling and Knox, in reading the landscape for tangible and evocative reminders of Australia's ancient and primitive past, Maloney and Walker mirrored the efforts of the nature writers of the 1920s and 1930s who articulated a valued past in particular natural sites and objects.[72] Unlike the nineteenth-century antiquarians and the nature writers of the twenties and thirties, however, and unlike Walling and Knox from the 1950s to the 1970s, who also sought to define a tangible cultural heritage in the vernacular rural landscapes of Australia's European pioneering past, Maloney and Walker repudiated the attitudes towards landscape of Australia's European ancestors, their 'strong European traditions', and need to control or struggle against nature and destroy 'much of the beauty inherent in the land'.[73] Such attitudes reflected the environmentalism of the 1960s and changing views of Australian natural landscape described by Robin.[74] Given

FIGURE 4.   *Banksias (black and white illustration) by Jean Walker. Source: Betty Maloney and Jean Walker,* More About Bush Gardens *(Sydney: Horwitz Publications Inc. Pty Ltd, 1967), pp. 74–75. © Jean Walker. Reproduced by permission of Elizabeth T. Smith and Ian and Rohan Walker, on behalf of Jean Walker. Permission to reuse must be obtained from the rightsholder.*

their negative perception of European Australia's human past, its removal from the recent past was thus central to Maloney and Walker's use of the continent's ancient and primitive past. The notion of plants as 'living fossils' established the middle-ground between past and present, bestowing relevance on this notion of past in the present.[75] The authority of science — not yet subject to the revision which has challenged the notion of scientific 'facts' as unproblematic — legitimized and perhaps encouraged the process.

Although they did not work together, Sydney-based landscape designer Bruce Mackenzie (b. 1932) explicitly identified with the ambitions and philosophy expressed in Maloney and Walker's two books about bush gardens, citing them as 'major influences'.[76] Mackenzie began practising as a landscape consultant in Sydney in the mid-1960s and played an active foundational role in the Australian Institute of Landscape Architecture (AILA).[77] He was also an outspoken advocate for the application of indigenous planting themes to the urban designed landscape.[78] Like Walling, Stones, Knox, Maloney, and Walker, Mackenzie also turned to the perceived naturalness of the Australian environment in attempts to develop and articulate an identifiably national cultural expression for the designed landscapes of Australia's cities and urban environments. He drew on the past, however, in a slightly different way from most of the other authors. Attributes of 'primitive' landscapes are inferred in Mackenzie's embrace of the 'mood experience' of 'remote natural landscapes' but, reflecting an environmental sensibility which he shared with Maloney and Walker, he was equally concerned about bushland sites in suburbia.[79]

Mackenzie justified his promotion of indigenous landscape themes by referring to them as the means 'for establishing a *tradition*' in Australian landscape design (my italics).[80] His use of 'tradition' to define a new approach to landscape design appears somewhat contradictory. But Mackenzie was acutely aware that use of Australian native plants and indigenous planting themes in landscape design, through the 1970s and even into the mid-1980s, amongst his own professional colleagues, had acquired the negative baggage of the 'unworldly', the purist, and the 'unsophisticated'.[81] Against this backdrop, Mackenzie's use of 'tradition' therefore appears strategic, to help counter resistance to his ideas by giving them 'the sanction of precedent'.[82] As Eric Hobsbawm argues, 'tradition' automatically implies 'continuity with the past'[83] and is a valued attribute of mature societies. According to Hobsbawm, traditions are also invented in contexts of 'desired change (or resistance to

innovation)', because they provide contrast in contexts of 'constant change and innovation of the modern world'.[84] Knowledge of the context in which Mackenzie deployed the notion of 'tradition' suggests that, consciously or unconsciously, both of these functions of tradition came into play augmenting his ideas with the gravitas of the past.

Conclusion

While there were differences and some contradictions both in the words used and the conceptions and depths of the past invoked, a striking and unifying factor in the examples from garden and landscape design literature espousing new design concepts is the gesture of reaching back. In doing so, the individuals who were conceiving of radically new kinds of Australian plant gardens worked with and against long-established imagery and meanings associated with Australian landscape and myths of national identity. What differed were the ways in which they invoked the past to justify the use of natural landscape and its manipulation to create the basis for a new and distinctly Australian ethos for landscape design.

A theoretical perspective helps to illuminate the significance of this gesture. In older, more established, cultures a sense of national identity and national greatness has often been built upon past cultural achievement, or 'golden ages'.[85] Lowenthal has applied similar thinking to new world contexts. In the absence of apparent or actual ethnicity and a lack of confidence in local cultural endeavour — common to new settler societies like Australia or the USA — aspects of natural landscapes are turned to 'to compensate for relatively recent human histories'.[86] It needs to be noted here that although writers like Knox and his contemporaries expressed sympathetic views towards the Australian Aboriginal people, the prevailing view at that time was that the culture of Aboriginal Australians was primitive, lacking sophistication, and static. This view shared much with thinking through the first half of the twentieth century about Aboriginal culture as confined to the past, and Aboriginal people as doomed to extinction.[87]

The process of establishing a deep sense of the past manifest in Australian plants and indigenous landscape as alternative foundational myths reduced critical dependence on Australia's remote British heritage and traditions provided by conventional gardening patterns. It could be argued that this freed garden and landscape designers from the constraints of conventional inherited gardening traditions, clearing a space for the emergence of new conceptions of what a garden could look like and from which a distinct and sophisticated Australian vocabulary for landscape design could develop. The Australian natural garden was part of an embryonic Australian design ethos that was not considered to have found full expression until the late 1970s and early 1980s, in projects like Illoura Reserve, Sydney, by Bruce Mackenzie (commissioned 1970);[88] the sculpture garden at the Australian National Gallery in Canberra by Harry Howard and Associates (c. 1978–1982);[89] and the master-plan concept for Royal Park in Melbourne by Brian Stafford and Ronald Jones (1984).[90]

The experimental and imagined Australian natural gardens of the 1950s to the 1970s were spaces in which new understandings of national identity, a sense of place, and a sense of place-in-time could flourish. But the moment was limited. The historic land rights campaign lead by Torres Strait Islander Eddie Mabo for Indigenous Australians, which began in 1982 and resulted in the 1992 High Court ruling that revoked *terra nullius*, appropriately began to invite and demand questions, such as, most notably, 'whose place?'[91] The significance of the burgeoning Australian natural garden and the literature which explored and espoused this new design concept over three decades is how they bear witness to changing attitudes towards Australian landscape and a transitional period in the formulation of understandings of Australian identity. Approaching this analysis of the Australian natural garden over time through the prism of Pratt's 'contact zone' has helped to reveal these changing attitudes towards Australia's natural landscape and the shifting baselines from which understandings of Australian national identity were formulated.[92]

Acknowledgements

For critical engagement with the work presented in this article I am grateful to the insightful comments of the anonymous reviewers which have refined and greatly strengthened my argument, and to audiences at the 'Gardens at the frontier' symposium, Waikato University, Hamilton, New Zealand (January 2014) and the World Congress of Environmental History 2014, Guimarães, Portugal (July 2014). An earlier version of this article was the recipient of the 2013 Mike Smith Prize, awarded by the National Museum of Australia in partnership with the Australian Academy of Science.

Disclosure statement

No potential conflict of interest was reported by the author.

NOTES

1. Frank Zelko, 'The Brick Veneer Frontier', in Christof Mauch and Helmuth Trischler with Lawrence Culver, Shen Hou, and Katie Ritson (eds), *Making Tracks: Human and Environmental Histories* (Munich: Rachel Carson Centre Perspectives, 2013/5), pp. 29–32; Katie Holmes, 'Growing Australian Landscapes: The Use and Meanings of Native Plants in Twentieth Century Gardens', *Studies in the History of Gardens & Designed Landscapes*, xxxi/2, 2011, pp. 121–130; Kylie Mirmohamadi, 'Designing Bush Landscapes: History and Place in Eltham and Castlecrag', *Studies in the History of Gardens & Designed Landscapes*, xxxi/2, 2011, pp. 131–138; Richard Aitken, *The Garden of Ideas: Four Centuries of Australian Style* (Carlton: Miegunyah Press, 2010); Jane Mulcock, 'Planting Natives: Gardening and Belonging to Place in Perth, Western Australia', in Frank Vanclay, Matthew Higgins and Adam Blackshaw (eds), *Making Sense of Place: Exploring the Concept of Place Through Different Senses and Lenses* (Canberra: National Museum of Australia Press, 2008), pp. 183–190.

2. Zelko, 'The Brick Veneer Frontier', pp. 29–30.

3. Ibid.

4. Benedict Anderson, *Imagined Communities: Reflections of the Origin and Spread of Nationalism*, revised edition (London, New York: Verso, 1991 [1983]), pp. 4, 6; Anthony D. Smith, *National Identity* (London and New York: Penguin, 1991), p. 21.

5. Anne Helmreich, *The English Garden and National Identity: The Competing Styles of Garden Design, 1870–1914* (Cambridge: Cambridge University Press, 2002), p. 4.

6. Mary Louise Pratt, 'Arts of the Contact Zone', *Profession*, xci, 1991, p. 34.

7. See Smith, *National Identity*; David Kaplan, Peter Catterall and Elfie Rembold, 'Introduction to Special Issue: *National Identities* in Retrospect', *National Identities*, xiii/4, 2011, p. 326; and David Kaplan and Guntrum Herb, 'How Geography Shapes National Identities', *National Identities*, xiii/4, 2011, p. 349.

8. Some exceptions include Katie Holmes, Susan K. Martin and Kylie Mirmohamadi, *Reading the Garden: The Settlement of Australia* (Carlton, Victoria: Melbourne University Press, 2008); and Allaine Cerwonka, *Native to the Nation: Disciplining Landscapes and Bodies in Australia* (Minneapolis: University of Minnesota Press, 2004).

9. See, for example, Holmes, 'Growing Australian Landscapes', pp. 121–130, and Philip Goad, 'New Land, New Language: Shifting Grounds in Australian Attitudes to Landscape, Architecture, and Modernism', in Marc Treib (ed.), *The Architecture of Landscape, 1940–1960* (Philadelphia: University of Pennsylvania Press, 2002), pp. 238–269.

10. The theoretical perspective responds to Sverker Sörlin and Paul Warde's call for engagement with 'the reservoir of thinking available to environmental historians' from the humanities and social sciences. See Sverker Sörlin and Paul Warde, 'The Problem of the Problem of Environmental History: A Re-reading of the Field', *Environmental History*, xii/1, 2007, p. 121.

11. In drawing on the forms, abstracted planting structures, and aesthetic of natural environments — in many cases, too, from diverse kinds of actual and remembered natural and mythologized rural landscapes — the imagined gardens discussed in this article are referred to as *natural gardens*, in recognition of their parallels with the wider, international phenomenon of twentieth-century natural garden-design. See, in particular, Joachim Wolschke-Bulmahn (ed.), *Nature and Ideology: Natural Garden Design in the Twentieth Century* (Washington, DC: Dumbarton Oaks Research Library and Collection, 1997).

12. Holmes, Martin and Mirmohamadi, *Reading the Garden*, pp. 98–103.

13. See, for example, Thomas Shepherd, *Lectures on Landscape Gardening in Australia* (Sydney: William McGarvie, 1836); C. Bogue Luffman, *The Principles of Gardening for Australia* (Melbourne: The Book Lovers' Library, 1903). For Joseph Maiden (1859–1925), see Jodi Frawley, 'Botanical Knowledges, Settling Australia, Sydney Botanic Gardens, 1896–1924' (PhD diss., Department of History, School of Philosophical and Historical Inquiry, University of Sydney, 2009); and on the use and meanings of Australian plants in inter-war gardens see Holmes, 'Growing Australian Landscapes', pp. 121–130.

14. Richard White, *Inventing Australia: Images and Identity, 1688–1980* (Sydney: George Allen & Unwin, 1981), pp. 148–149.

15. Goad, 'New Land, New Language', pp. 239–240.

16. White, *Inventing Australia*, p. 85.

17. See, for example, Alistair Knox and Bruce Mackenzie, 'The Indigenous Environment as a Concept for Applied Landscape Design' in *Proceedings of the Conference The Landscape Architect and the Australian Environment, conducted by the Australian Institute of Landscape Architects, Prince Philip Theatre, University of Melbourne, 30 August 1969* (Canberra, Australian Institute of Landscape Architects, 1970), pp. 39–45 [Knox], 46–49 [Mackenzie]; p. 48.

18. Holmes, 'Growing Australian Landscapes', p. 127.

19. Encouragement of close observation spanned horticultural writing and landscape design discourses through the period examined in this account. Two examples of such encouragement, selected because they bookend the period considered here, are Thistle Harris, *Australian Plants for the Garden: A Handbook on the Cultivation of Australian Trees, Shrubs, Other Flowering Plants, and Ferns* (Sydney and London: Angus and Robertson, 1953), pp. 160–161, and Glen Wilson, 'Towards an Australian Style of Landscape Design', *Landscape Australia*, i, 1979, p. 42.

20. Edna Walling, *The Australian Roadside* (Melbourne: Oxford University Press, 1952).

21. Tom Griffiths, *Hunters and Collectors: The Antiquarian Imagination in Australia* (Melbourne: Cambridge University Press, 1996), pp. 141–149.

22. White, *Inventing Australia*, p. 85.

23. Ibid.

24. Catriona Elder, *Being Australian: Narratives of National Identity* (Crows Nest, NSW: Allen & Unwin, 2007), pp. 31–38.

25. Griffiths, *Hunters and Collectors*, p. 144.

26. Beaumaris Tree Preservation Society, *Native Plants for Seaside Gardens: Local Lore, Birds and Wildflowers* (Prahran: Fraser & Morphet, 1954), pp. 1, 4, 16.

27. Drew Hutton and Libby Connors, *A History of the Australian Environment Movement* (Cambridge: Cambridge University Press, 1999), pp. 96–97.

28. See, in particular, Walling's closing verse (unattributed) which states: 'The roadside is the Front Garden of the Nation' cited in Walling, *The Australian Roadside*, p. 110.

29. Edna Walling, 'Letters to the Editor: Un-Australian Tree Planting', *The Age* (24 November 1955), p. 2.

30. Libby Robin, 'Urbanising the Bush: Environmental Disputes and Australian National Identity' in David Dale (ed.), *Australian Identities* (Melbourne: Australian Scholarly Publishing, 1998), p. 116.

31. Libby Robin, *Defending the Little Desert: The Rise of Ecological Consciousness in Australia* (Carlton South, Victoria: Melbourne University Press, 1998).

32. David Lowenthal describes this kind of conception of the past as a valued attribute particular to primitiveness in *The Past is a Foreign Country* (Cambridge: Cambridge University Press, 1985), pp. 53–54.

33. Walling, *The Australian Roadside*, p. 17.

34. R. Wright, *The Bureaucrat's Domain: Space and the Public Interest in Victoria, 1836–84* (Melbourne: Oxford University Press, 1989); Georgina Whitehead, *Civilising the City: A History of Melbourne's Public Gardens* (Melbourne: City of Melbourne with the State Library of Victoria, 1997).

35. Anne Latreille, 'Stones, Ellis ('Rocky')' in Richard Aitken and Michael Looker (eds), *The Oxford Companion to Australian Gardens* (South Melbourne:

36. Collection of landscape design drawings by Ellis Stones [picture], YLTAD 35, State Library of Victoria [SLV]; and Architectural Drawings by Graeme Gunn, Architectural Drawings LTAD *189 Bag 14*, SLV.

37. Alistair Knox, *We Are What We Stand On: A Personal History of the Eltham Community* (Eltham: Adobe Press, 1980), p. 102.

38. Ellis Stones, *Australian Garden Design* (South Melbourne: Macmillan, 1971), pp. 64–65.

39. Lowenthal, *The Past is a Foreign Country*, p. 62.

40. Ibid., pp. 59–62.

41. The community at Eltham and Alistair Knox have been introduced previously in this journal by Kylie Mirmohamadi, 'Designing Bush Landscapes'.

42. Knox's practice as an environmental designer commenced in c. 1945. He articulated his vision and philosophy for design in harmony with the environment in 1975 in Alistair Knox, *Living in the Environment* (Canterbury, Victoria: Mullaya Publications, 1975) and in 1980 in *We Are What We Stand On*.

43. Alistair Knox, 'Notes for meeting held at Mount Pleasant Road on 15 February 1968 to discuss formation of "Eltham Preservation Society"', Peter Glass Papers, PA 99/87, Australian Manuscripts Collection, SLV.

44. Knox and Mackenzie, 'The Indigenous Environment'.

45. Knox, 'The Indigenous Environment', pp. 40, 42.

46. See, in particular, the chapter 'The Australian Identity' in Knox, *Living in the Environment*, pp. 92–96.

47. J. B. Hirst, 'The Pioneer Legend', *Historical Studies*, xviii/71, 1978, pp. 316–337.

48. Knox, 'The Indigenous Environment', pp. 44–45.

49. Knox, *Living in the Environment*, p. 92. The promulgation of the peaceful settlement doctrine by Knox reflected widely held beliefs contemporary with the period discussed, which have begun to be challenged by scholars such as Henry Reynolds in *The Forgotten War* (Sydney: NewSouth Publishing, 2013), p. 16.

50. Lowenthal, *The Past is a Foreign Country*, p. 63.

51. Knox, 'The Indigenous Environment', pp. 42, 45.

52. Betty Maloney and Jean Walker, *Designing Australian Bush Gardens* (Sydney: Horwitz Publications, 1966).

53. Reviews for *Designing Australian Bush Gardens* appeared in the major daily papers in Sydney and Melbourne (*Sydney Morning Herald* and *The Age*), booksellers' lists, popular gardening magazine *Your Garden*, and journals of the Society for Growing Australian Plants [SGAP] and the Rangers League of NSW. The review cited was from agricultural newspaper *The Land*. See 'What the Critics Said About … ' in Betty Maloney and Jean Walker, *More About Bush Gardens* (North Sydney: Horwitz Publications, 1967), p. 125.

54. Robin, 'Urbanising the Bush', p. 116.

55. Maloney and Walker, *More About Bush Gardens*, p. 3.

56. Similar kinds of references appear at multiple points in the text of *More About Bush Gardens*. Ferns are featured on pp. 14–23, banksias on pp. 70–79.

57. Maloney and Walker, *More About Bush Gardens*, p. 36.

58. Ibid., p. 11.

59. Joseph Dalton Hooker, *The Botany of the Antarctic Voyage of H. M. Discovery Ships Erebus and Terror, In the Years 1839–1843, Under the Command of Captain Sir James Clark Ross, KT., R. N., F. R. S. & L. S., etc. Part III. Flora Tasmaniæ, Vol. I. Dicotyledones* (London: Lovell Reeve, 1860), pp. xxii, xxvii; Professor A. C. Seward, ScD, FRS, 'The Vegetation of Gondwana Land', *Report of the Eighty-Fourth Meeting of the British Association for the Advancement of Science, Australia: 1914, July 28–August 31* (London: John Murray, 1915), p. 584.

60. Rachel Sanderson, 'Re-writing the History of Australian Tropical Rainforests: "Alien Invasives" or "Ancient Indigenes"', *Environment and History*, xiv, 2008, pp. 165–185. See also H. E. Le Grand, *Drifting Continents and Shifting Theories: The Modern Revolution in Geology and Scientific Change* (Melbourne: Cambridge University Press, 1988).

61. See, for example, Frederick Chapman, 'Excursion to the Botanical Gardens', *Victorian Naturalist*, xlvii/1, 1930, p. 19.

62. Alan Bell, 'In the Footsteps of Edward John Eyre: Narrative of the Grimwade Expedition', *Wild Life*, six part series from January to June 1948; also *Wildlife and Outdoors*, March 1953, pp. 254, 257.

63. Thistle Y. Harris, *Wild Flowers of Australia* (Sydney: Angus & Robertson, 1939). An enlarged and updated version of *Wild Flowers* was published in 1947, and reprinted 1948, 1952, 1956, 1962, 1966, 1971, 1979; see also Thistle Harris, Interviewed by Hazel de Berg in the Hazel de Berg collection (7 January 1972), [sound recording], National Library of Australia.

64. Harris, *Wild Flowers of Australia*, p. xiii.

65. *Report of the Eighty-Fourth Meeting of the British Association for the Advancement of Science*, p. 138.

66. Thistle Y. Stead Harris, 'Letters: Edna Walling Remembered', *Landscape Australia*, i, 1986, p. 10.

67. Jean Walker, Interviewed by Burge/Read (Interview 1), 19 August 2007, Australian Garden History Society, Sydney and Northern NSW Branch Oral history Project — Miss Jean Walker (neé Brown) [transcript].

68. Elyne Mitchell and Harold Cazneaux, *Australian Treescapes: A Photographic Study* (Sydney: Ure Smith, 1950); and James Wales Audas, *The Australian Bushland* (North Melbourne, Victoria: WA Hamer, 1950).

69. Elyne Mitchell, 'Australian Treescapes', in Mitchell and Cazneaux, *Australian Treescapes*, p. 11.

70. George Althofer, *The Story of Nindethana and New and Enlarged Catalogue of Australian Native Plants* (Dripstone, NSW: Nindethana Nursery, 1956), p. 78.

71. Lowenthal, *The Past is a Foreign Country*, pp. 52–57, 61.

72. Griffiths, *Hunters and Collectors*, p. 144.

73. Maloney and Walker, *Designing Australian Bush Gardens*, p. 9.

74. Robin, 'Urbanising the Bush', pp. 116–117.

75. Lowenthal, *The Past is a Foreign Country*, p. 62.

76. Andrew Saniga, *Making Landscape Architecture in Australia* (Sydney: UNSW Press, 2012), p. 71.

77. Mackenzie was a member of the AILA Federal Council (by 1971) and served as State President (1980–1981) and National President (1981–1983). 'Foreword' in *Landscape Australia*, 1, June 1971, p. 2. AILA Archive, courtesy of Andrew Saniga; and www.aila.org.au/profiles/mackenzie/default.htm.

78. Bruce Mackenzie, 'The Landscape Environment: A Wasted Potential', *Architecture in Australia*, lv/6, 1966, pp. 111–120; and Knox and Mackenzie, 'The Indigenous Environment', pp. 47–49.

79. Mackenzie, 'The Landscape Environment', pp. 114–115.

80. Mackenzie, 'The Indigenous Environment', p. 48.

81. Bruce Mackenzie, 'Nothing More Relevant Than Relevance: Part 2 of Artistry, Relevance and the Landscape Architect', *Landscape Australia*, i, 1986, p. 31.

82. Eric Hobsbawm, 'Inventing Traditions' in Eric Hobsbawm and Terence Ranger (eds), *The Invention of Tradition* (Cambridge: Cambridge University Press, 1983), p. 2.

83. Ibid., p. 1.

84. Ibid., p. 2.

85. Edward Said, *Culture and Imperialism* (New York: Knopf, 1993), p. xiii.

86. Lowenthal, *The Past is a Foreign Country*, p. 54.

87. Griffiths, *Hunters and Collectors*, p. 55.

88. Bruce Mackenzie, 'Alternative Parkland', *Landscape Australia*, i, 1979, p. 19. Illoura Reserve is classified as significant at a State level (to NSW) in January 2014, as 'an outstanding twentieth century urban park that was an important forerunner to the implementation of the Sydney Bush School landscape design philosophy in public parks'. http://www.environment. nsw.gov.au/heritageapp/ViewHeritageItemDetails. aspx?ID = 5061932.

89. Harry Howard, 'Landscaping the High Court of Australia and the Australian National Gallery — The Sculpture Gardens', *Landscape Australia*, iii, 1982, pp. 208–215.

90. 'Royal Park is, along with the Harry Howard and Associates' Sculpture Garden at the National Gallery in Canberra, widely considered the most important Australian landscape design of its time.' See editors' note to Brian Stafford and Ron Jones, 'Royal Park Melbourne', in Harriet Edquist and Vanessa Bird (eds), *The Culture of Landscape Architecture* (Melbourne: Edge Publishing Committee, 1994), p. 167.

91. George Seddon, *Swan Song* (Centre for Studies in Australian Literature, University of Western Australia, Perth, n.d.), pp. 14, 211, cited by Tim Bonyhady and Tom Griffiths, 'Landscape and Language', in Tim Bonyhady and Tom Griffiths (eds), *Words for Country: Landscape and language in Australia* (Sydney: University of New South Wales Press, 2002), p. 10.

92. New Zealand-based landscape architect Shelley Egoz has explored the tension that exists when identity is drawn from landscape, when both identity making and landscape, inherently, are in continual flux. See Shelley Egoz, 'Landscape and Identity: Beyond a Geography of One Place' in Peter Howard, Ian Thompson and Emma Waterton (eds), *The Routledge Companion to Landscape Studies* (London: Routledge, 2013), in particular pp. 275–276.

W. W. Smith and the transformation of the Ashburton Domain 'from a wilderness into a beauty spot', 1894 to 1904

MICHAEL M. ROCHE

Introduction

Born in Hawick in Roxburghshire, William Walter (but only ever W. W.) Smith (1852–1942) served his five-year gardening apprenticeship locally at Wilton Lodge, Hawick (1866–1870), before moving on to work in the Lake District at Rosehill and Lowithwaite Hall in Keswick. It seems reasonable to assume that he was familiar with Scottish formal gardens and with the extension of romantic ideals and the creation of ordered park lands that spread into Scotland in the late eighteenth century after the style of the famous English landscape architect, Lancelot 'Capability' Brown. In the 1870s, Robinson's ideas about 'natural gardens' were also finding some acceptance in Scotland, so that Smith was potentially exposed to a variety of influences during his apprenticeship.[1] This would seem to be borne out in testimonials that stress his broad-based knowledge and competence. From 1871, he served three-and-a-half years as under-gardener and later foreman, at the famous Burghley Estate, near Stamford in Lincolnshire, England. William Temple, Burghley's head gardener, described him as 'strictly honest, steady and truthful', with a thorough knowledge, as well as being 'skilled in practical geometry and in the laying out of grounds and in every sense of the word a very intelligent young man'.[2]

Burghley Estate was notable for being redesigned by 'Capability' Brown, from 1755 to 1779.[3] During this time, Brown altered the façade of the house, created a new lake, erected a bridge, constructed a hill, established new sweeping lawns and planted many trees, while removing others.[4] Smith worked at Burghley a century on from Brown's design work and must have been conscious of the latter's importance as a landscape gardener. At a day-to-day level he would have been involved in maintaining some of the park features first designed by Brown albeit with later modifications.

From Burghley, Smith migrated to New Zealand, settling in Canterbury and finding employment as a gardener on one of the pastoral estates (figure 1). Thelma Strongman and Katharine Raine have provided separate overviews of the types of gardens in New Zealand from the late Victorian and Edwardian eras.[5] In New Zealand, Raine identified geometrical flower beds and planting out as an 'enduring feature', along with a mix of formalistic and more naturalistic approaches that represented a relaxation of 'the strictures of the conventional British gardens'.[6] Grass, she noted, remained a 'characteristic feature of the Victorian garden', in addition to the small cluster of native cabbage trees (*Cordyline australis*), flax (*Phormium tenax*), and tree ferns (*Cyathea* and *Dicksonia* spp.) or toetoe (*Cortaderia* spp.).[7] Strongman focused specifically on Canterbury, with its greater temperature extremes and lower rainfall than the North Island as well as its legacy as a planned settlement. She described six elements of Edwardian gardens in Canterbury, these being: (1) their rose garden; (2) their native garden; (3) the rockery; (4) flowering trees and shrub borders; (5) bog gardens and water gardens; and (6) subtropical

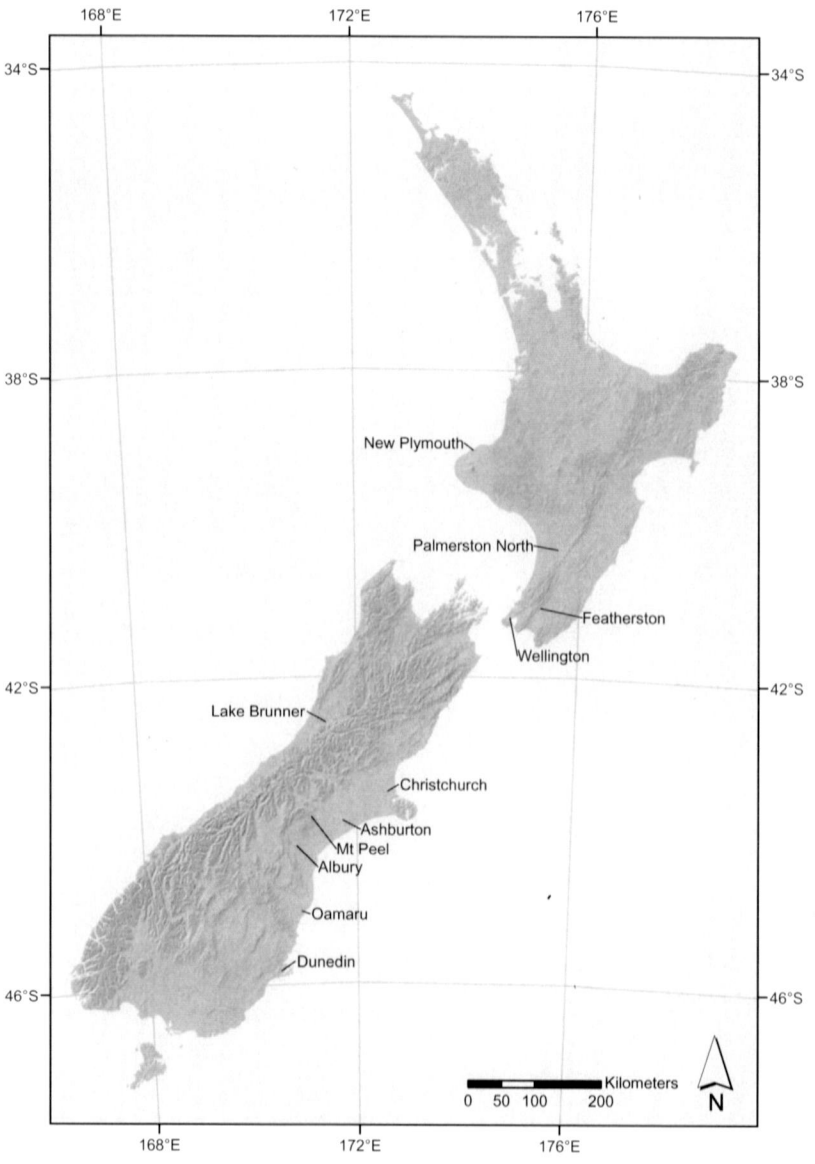

FIGURE I. *New Zealand. Location of main places of significance to Smith that are mentioned in the text. Source: Map drawn by Alistair Clement.*

gardens. The rose she describes as a plant 'held in the greatest esteem' in the earliest decades of the twentieth century, noting also a tendency for these to be grown typically in 'geometrically shaped beds and on pergolas' in a rosary rather than in a more naturalistic manner.[8] Both Strongman and Raine share a concern for private gardens, though not only of the wealthy in the case of the latter. Smith was, however, curator of a public garden as well as straddling the Victorian and Edwardian eras and this is reflected to some extent in the assemblage of features he accentuated and developed in the Ashburton Domain.

The term domain here referred to an open, typically urban space with gardens, lawns, plantings of trees, sometimes also including sports grounds for recreation and beautification. The New Zealand literature on urban parks and domains is unevenly developed. Some of the major botanic gardens have been the subject of individual historical treatment and there has been at least one national comparative account of parks and domains, but there has only been fragmented scholarly treatment.[9] The paper also adds to the small stock of biographical studies of New Zealand park and domain curators and landscape architects.[10]

With respect to the theme of 'gardens at the frontier', W. W. Smith's time at the Ashburton Domain accentuates three points, the first of which is that the 'frontier' that it represented was not one principally defined by location. The pioneering pastoral fringe in Mid-Canterbury, such as it was, dated back to the early 1850s, whereas Smith was at work half a century later, almost two generations further on. Second, Smith's 'frontier' was rather one that involved transforming a 'wilderness' into an aesthetically pleasing environment for passive recreation, contemplation and education (along with some more active sports). This making over involved filling a 'wilderness' regarded as an empty space, a barren void rather than distant untouched pristine nature, with gardens of flowers, shrubs, trees and lawns. Third, and finally, as curator, Smith appeared at a stage in the Domain's history where there had been several decades of piecemeal development and he was unafraid to apply his training in reshaping the Domain in some quite dramatic ways. Notably, he was prepared to remove trees planted in earlier years, to create what locals regarded as a beauty spot and also to introduce some native species into the Domain. This may be interpreted as the taming or domestication of selected indigenous species and inter-mixing them with European species, though subsequently, it is argued, that indigenous species became increasingly important to Smith's aesthetic in their own right.

Early experiences in New Zealand

Around 1875, Smith arrived in New Zealand, and worked industriously for four-and-a-half years as gardener on John Acland's Mt Peel Station in Canterbury, a notable colonial estate with its large garden and extensive plantings of Australian eucalypts, conifers, elms, oaks and cedars.[11] The sheep station eventually boasted a sizable vegetable garden and orchards, plus a pleasure garden and fernery. Smith was employed at a modest £60 p. a., Acland was a keen gardener and, as Strongman observed, 'a head gardener could probably never feel completely in charge while his employer might suddenly begin to work by his side'.[12] The details and events of Smith's work at Mt Peel remain elusive, but there are glimpses; during his tenure, the pleasure garden was developed with rhododendrons from Christchurch in 1877 and camellias from the UK in 1878, and in 1878 the deodar produced its first cones.[13] He is also credited with planting the first giant Himalayan lilies (*Cardiocrinum giganteum*) at Mt Peel. Seed labelled as 'Mr Smith's Lilies' is still available at the church at Mt Peel.[14] Smith stayed until 1880, resigning so that he could get married.

Thereafter he was employed as gardener at the nearby Albury Estate. This was owned by Edward Richardson (1830/31–1915), an English-born and trained engineer who arrived in Christchurch via Australia. He had secured a lucrative contract to construct a railway and tunnel under the Port Hills linking Christchurch with Lyttelton, which was completed in 1867. Richardson then purchased Albury Estate, which he rapidly placed in freehold and expanded while immersing himself in provincial politics. He was an enthusiastic tree-planter and had over 100 acres in such by 1882, including redwoods and Oregon pines.[15] In 1883, Richardson sold off 3400 acres of Albury for closer land settlement and Smith appears to have left his employment around this time.[16]

From here, Smith was gardener at Edward Menloe's Windsor Park, near Oamaru, North Otago, where he remained until around 1886. Menloe's efforts in tree-planting and estate improvement were acknowledged locally, though he enjoyed a greater reputation as a stock breeder. While at Windsor Park, Smith won prizes for his grapes and roses in the 'open' as opposed to 'amateur' section of the local North Otago Horticultural Society competitions, but in later years, here and elsewhere, he was more often a judge than a competitor.[17] Thereafter, Smith's movements and occupations, for a time, are more difficult to track. He mentioned spending 1887 and 1888 working near Lake Brunner in Westland.[18] He elsewhere noted that he was in Australia in 1889, but seems to have returned to New Zealand immediately afterwards.[19] There are some suggestions that he may have been working as a freelance gardener in North Otago, but by 1890 he was residing in Ashburton, Mid-Canterbury.[20]

In 1894, he was the successful applicant, out of 22, for the position of Caretaker of the Ashburton Domain.[21] This appointment offered some security to Smith, who by this stage had a young family to support. The salary was hardly generous at £78 p. a., but a Caretaker's Lodge came with the position.

Becoming a natural historian

Smith's reference from Burghley Park highlighted his intelligence and interests. These were expressed in his wide-ranging engagement with natural history, especially entomology and ornithology, which later expanded to include archaeology and ethnography. While employed at Mt Peel Smith located the then rare and now extinct Laughing Owl (*Sceloglaux albifacies*) nearby and later captured some alive for the noted ornithologist Walter Buller.[22] In this role as a collector, Smith corresponded with many nationally important figures in natural history in New Zealand, including Buller, Julius von Haast, and George M. Thomson. Although he was primarily a keen field observer, his diligent note-taking meant that he was especially useful at recording the particular behaviour of different species and at noting the first occurrence of native and introduced species in an area. Some of Smith's observations were sufficiently worthy for high-profile naturalists, such as Buller and Thomson, to read them as papers on his behalf to the Wellington and Otago philosophical societies, branches of the New Zealand Institute.[23] From modest beginnings as a gleaner who donated collections of New Zealand fossils and insects to the Otago Museum while at Windsor Park,[24] he eventually produced more substantial accounts, for instance a lengthy paper on the avifauna of Lake Brunner district published in the *Transactions and Proceedings of the New Zealand*

Institute, as well as contributions to British journals such as *Ibis*, the *Entomologists' Monthly Magazine* and the *Entomologist*.[25] He also participated in a detailed newspaper debate about the extinction of the moa in which he logically deconstructed and rejected some of the more fanciful theories of the time.[26]

An interest in the moa perhaps triggered and certainly coincided with Smith's growing curiosity about the 'ancient' Maori past, then understood in terms of the arrival of a 'great fleet' around AD 1350.[27] He was a foundation member of the Polynesian Society, established in 1892 by Stephenson Percy Smith and Edward Tregear, although some of his writings about stone implements and rock drawings were published much later, and only after he had taken up his position in Ashburton.[28] Historian of science Ross Galbreath noted that Smith's attitudes changed over time and that from having been an avid bird collector in the 1880s, by 1895 he was writing publically in support of the government's recent setting aside of Little Barrier and Resolution Islands as sanctuaries.[29] His support for native bird protection extended subsequently to forests. In 1903, he corresponded with Percy Smith about the damage being done to state forests, urging government action to conserve them for utilitarian and scenic purposes. This connection would be important for the direction of Smith's later career.[30]

This large volume of writing and associated correspondence relating to natural history reveals where Smith's interests really lay and perhaps explains in part why he never seems to have expounded upon his ideas about gardening, horticulture or landscape design in print. This contrasts with another public gardener, David Tannock, Superintendent of the Dunedin City Botanical Garden and City Reserves, who produced the first edition of his *Manual of New Zealand Gardening* in 1914.[31] Certainly Smith's surviving personal papers held in Puke Ariki, New Plymouth's museum, apart from a small amount of family material, deal almost entirely with various natural-history topics.

Some features of the Ashburton Domain that date back to Smith's time are still, however, evident in its present landscape, over a century on from when he fashioned them. It is possible to discern them — and hence his landscaping — by bringing together period minute books, newspaper accounts, and pictorial evidence, in the form of postcards, in conjunction with a number of site visits to the Domain. This has not been a straightforward task, for a severe windstorm in 1975 toppled many of the older and taller trees and others have been removed subsequently as part of ongoing developments, meaning that many of Smith's contributions have been erased from the landscape.

Curator of Ashburton Domain 1894 to 1904

Ashburton is a country town centrally located on the Canterbury Plains approximately equidistant between the cities of Christchurch and Timaru. The dry and exposed tussock-grassland plains were rapidly taken up as large pastoral estates in the early 1850s before a town was laid out by the ford across the Ashburton River. Town and river were both named after one of the principal figures in the Canterbury Association, which had established a colony, as a planned settlement along the lines advocated by Edward Gibbon Wakefield, in Canterbury in 1850. By 1901, the borough population had reached 2082 and the town as a rural service centre was benefitting from the frozen meat trade to the UK initiated in the early 1880s.

A domain reserve was part of the original Ashburton town plan, which was laid out in 1861, although a Domain Board was not formed until 1874. Located immediately inside the town belt on the north-eastern side of the settlement, the area was bounded on the town side by an old river channel. In 1880, water was diverted into this from the nearby Mill Creek to form what was described as 'one of the prettiest artificial lakes in the colony'.[32] This included a horseshoe bend and another larger lake with an island. Control of the 90-acre domain was vested in the local Borough Council in 1889, after which a swimming pool and cycling track were constructed.[33]

In the absence of any material in Smith's personal papers on landscape aesthetics and horticultural techniques or any expressions of other utilitarian considerations about developing a small-town park, this paper has necessarily had to rely on other sources. The Domain Board Minute Books, along with the summaries of these meetings in the local newspaper and other extraneous newspaper items provide an outline of Smith's work in the Domain. Viewed across a decade, they reveal the breadth of his activities and some of the difficulties he encountered, not only those imposed by financial limitations, but also by frost, drought and storm damage.

Smith was not the first caretaker appointed to the Domain, but arguably by dint of his formal training, vision and effort he was one of the most important appointees, responsible for shaping and reshaping the Domain's landscape.[34] Smith's skill and efforts soon saw him designated curator and his salary was raised from £78 to £100.[35] Whereas at Mt Peel Acland took a personal interest in his garden and doubtless closely directed Smith, the situation in Ashburton was different. For the first time he was responsible to a public board and one that was not financially well provided for. For the first time, too, he was subjected to decision-making by committee, with all the tensions typically associated with a small country town. But, more than this, it meant that Smith had to provide a fortnightly written report to the Domain Board, as well as needing authorization to undertake even some quite minor tasks and implement board decisions. Unfortunately for this present study only a few, not especially detailed, examples of Smith's fortnightly reports survive. In addition, the minute books typically note only that his reports were received and accepted, so that a sustained first-hand record of his work and plans does not appear to have survived.

A public domain also generated other tasks, that, if not exactly onerous to Smith, were time consuming and recurring, such as dealing with cases of trespass involving people and dogs, seeking prosecution of individuals killing ducks, and attending to cases of 'furious riding' by cyclists on the paths through the Domain. Fortunately only occasionally did he have to herd escaped bulls and 'wild cattle' back into the Domain paddocks. These were leased out by the Board for grazing and thus a welcome additional source of revenue. The Domain was not exclusively given over to parkland and gardens, but included tennis courts, a cricket pitch and, at that time, hockey grounds. This was a further source of friction. Smith was both critical of the Sports Association and criticized by them over damage to and preparation of the grounds. That the Domain Board persisted for a time in employing a separate sports ground curator may not have helped matters.

The Domain was still fairly unfinished when Smith was appointed, so that he was also able to stamp his mark on it in the longer term. That said, he faced some formidable difficulties, mostly stemming from the lack of funds available for development. Smith's initial work included a first thinning out of trees planted years earlier, along with improvements to the tracks through the Domain, and expansion of the flower beds. This was heavy work with the old stumps. Broom, couch grass and cocksfoot grass removed in this process were burnt and the ash spread over the new beds as fertilizer.[36] This first large-scale establishment of flower beds was another initiative on Smith's part.[37] One prolonged source of frustration was the difficulties encountered in purchasing a suitable lawnmower and obtaining replacement parts. This also signals, consistent with Raine's observations, the importance of grass to the look of the Domain as a whole, and Smith's desire to create picturesque scenes, over and above the sometimes fraught requirements of cricket-ground preparation.

Smith was actively involved in tree planting, but the Domain in the early years of his tenure suffered from some unseasonal climatic and weather events. From 1896 to 1899 a drought damaged many of the trees, while in 1898 exceptionally heavy frosts killed the Australian wattle and *Eucalyptus* trees.[38] Later, in 1898, Smith made significant plantings of various *Pinus* species. Subsequently, he labelled trees with their botanical names in order to help educate the general public.[39]

Although the Domain Board granted him 10/- to purchase seeds and flowers in 1898 and authorized the purchase of 150 trees in 1900, he was heavily dependent on donations of plants and seeds, which must have considerably restricted his efforts even allowing for the fact that he was raising his own plants from seed. Borough Councillor Hugo Friedlander gifted dahlias while other local notables, such as W. H. Collins (a future mayor), David Thomas and a Mr Cochrane, also offered plants.[40] The majority of the gifts were of unspecified bulbs and seeds, but included

TABLE I.   *Donations of plants, seeds and trees to the Ashburton Domain Board 1894–1904.*

Year	No. of separate donations	Roses	Dahlias	Lilies	Natives
1894	I				
1895	5		12		
1896	14	I lot of 12 (all named)			2 lots
1897	14	2 lots		I	2 lots
1898	9				I
1899	6				
1900	4				
1901	5				
1902	5				I
1903	—				
1904	—				

Source: Compiled from Ashburton Domain Board Minute Books numbers 3 and 4.

FIGURE 2. *Rose garden and paths created by W. W. Smith in the Ashburton Domain. Source: Author's collection, F. T. [Frederick Tanner] Glossine Series, postmarked 1908. © Author's collection. Reproduced by permission of the author. Permission to reuse must be obtained from the rightsholder.*

FIGURE 3. *Rose garden, Ashburton Domain in 2013 showing the rose garden and paths originally laid out by W. W. Smith. This garden has been considerably modified in 2014 most notably by the addition of a pergola. Source: photograph by author.*

quantities of fashionable roses, grass seed, and a few select consignments of native ferns, and other native plants including cabbage trees (table 1).

Two examples of elements of Smith's work that remain in the present Domain can be traced via period postcards, verified by site visits (see also the last section). William Temple commented on Smith's 'skill in geometry' and it is possible to see an example of this in a period postcard (figure 2). This scene primarily shows the bowling green, which was created in 1898. Nearby, precisely cut into the turf, is a star-shaped bed planted with herbaceous plants. Beside this, there is circular rose bed surrounded by a path, with other pathways leading to and from it and stone stairs providing access to the bowling green behind.[41] This nicely reinforces Stongman's view of the importance of roses in this period. In 1895, with the assistance of two wage workers, Smith had completed the rose beds, which involved much moving of soil by wheelbarrow. A dozen named roses had been donated in 1896 and are likely those in

the photograph. This rose garden still exists today, albeit in a somewhat modified form (figure 3).

In other parts of the Domain, Smith concentrated on trees and instead of straight pathways favoured curvilinear paths and more naturalistic plantings. These are shown in a period postcard of the gates erected on the occasion of Queen Victoria's jubilee in 1897 (figure 4). The Jubilee Gates were flanked by oak trees provided by the Railways Department, their yard and station being situated across the road from the gate, which was also adjacent to the curator's lodge, long-since demolished. Again, this sweeping pathway created by Smith remains part of the Domain to the present, though the maturation of trees and later plantings make it impossible to re-photograph the scene from the same angle. The

In the Ashburton Domain.

F. T. Series. No 112 A.

FIGURE 4. *Smith's curved pathway and more naturalistic plantings including shrubs and trees leading to the Queen Victoria Jubilee Gates, which are just visible behind the man's head. The figure is not Smith. There is a New Zealand cabbage tree (Cordyline australis) in the left foreground. Source: Ashburton Museum Collection, F. T. [Frederick Tanner] Series No. 112 A. © Ashburton Museum. Reproduced by permission of Ashburton Museum Collection. Permission to reuse must be obtained from the rightsholder.*

appearance of this path and associated beds upholds Raine's comments about the relaxation of formal design elements in New Zealand gardens at this time.

The stocking of the Domain was also undertaken with a range of bird and fish species, some donated. These included Muscovy ducks in 1895, frogs (1896), two peafowls (1897), paradise ducks (1900), and a pair of ducks (1900).[42] Trout were released into the ponds in 1901 and the *Cyclopaedia of New Zealand* reported the presence of black and white swans, perch and other fish, and hundreds of native ducks during the shooting season.[43] The swans, both black and white, also feature in many of the period postcards of the Domain. The Domain at this point was something of a menagerie, not solely a public garden of plants fulfilling a subsidiary function of botanical education. The extent to which it was a sanctuary for avifauna sets it apart from the private gardens of Canterbury of the time. In 1903, Smith began work on

trimming the undergrowth on the island in the main pond and replanting it with native trees and shrubs to create a comparatively large 'native garden'. This work has continued to the present day.[44] Later in 1903, as winter approached, he shifted his efforts onto major track-widening to accommodate both pedestrians and cyclists.[45]

Beyond his duties as curator, Smith busied himself in the local community serving as judge for the local horticultural society and as a committee member of the local beautifying society, in addition to assisting the local Agricultural & Pastoral Association with the identification of farm pests and publishing on the natural history of the district.[46] As well as writing about Maori artefacts, he contributed a large number of shorter notes to the *Entomologists' Monthly Magazine* and the *Entomologist*.

For his efforts in developing the Domain, Smith was publically lauded as 'a man of exceptional taste and resource'.[47] Smith, however, did not write about the sorts of principles that guided his work as a gardener, so that while some of his activities can be reconstructed, his motivations can only be examined indirectly. Others in the town were willing to fill this gap, and although none of them might be regarded as his peers, it is clear that Smith's efforts were popular and well regarded by the standards of the day. In 1896, the *Ashburton Guardian* published a lengthy illustrated article by 'Starling Gray' that quoted from poets Tennyson and Cowper, then claimed that 'an hour spent in the Domain is time well spent and will repay both the seeker after sylvan pleasure and the student of Nature'.[48] Gray declared that, twenty years earlier, the Domain had been 'a wild unkempt tussock field, and in memory a gaze at the desolate scene it presented, and to compare it with that scene of today is a pleasing mental effort'.[49] While he acknowledged the contribution of earlier individuals, Gray was in no doubt that Smith contributed markedly to 'the creation of its beauty and its value as a botanical garden'. Alluding to Smith's training, and reinforcing the remarks made by his superior at Burghley, Gray described him as:

Well informed, thoroughly master of gardening both as a science and an art, whether it be in landscape gardening, in the more limited area of floriculture, or in [the] wider range that arboriculture affords, Mr Smith has been the great success he was expected to be.[50]

Gray lauded Smith as a 'man of exceptional taste and resource'.[51] Other praise appeared only after his resignation was announced in 1904. 'Scenery', in letters to the editor, stated that 'the work performed by him [Smith] in our public gardens will be a lasting monument if one is required to his energy and hard work'.[52] A subsequent editorial noted that although he was not one of the founders of the Domain, 'it is indisputable that much of its present beauty is due to his careful solicitude and diligent application'.[53]

Further public appreciation followed later in 1904. One letter-writer, a long-term resident ventured, as 'one who has seen the Ashburton domain grow from a wilderness into a beauty spot I feel sorry and glad we are losing Mr W. W. Smith. The work performed by him will be a lasting monument, if one is required, of his energy and hard work'. (The pleasure came from the thought that Smith's skills would be made available to those in other parts of the country.)[54] The clearest expression of Smith's landscape aesthetic emerges tangentially in a letter-to-the-editor in the *Ashburton Guardian* challenging decisions made by the Domain Board and the efforts of James Young, Smiths' successor as curator. 'Rate Payer by the Dozen' claimed:

> There are in the domain a hundred and one landscape pictures, and a man with a truly artistic eye and keen conception can make and keep each beautiful. The whole when thus kept, make the domain the 'beauty spot' we all wish it to be. Our old W. W. gave his chief attention to these pictures (where is his island picture now? Where are the lordly pines that Smith left?) his eye was of Nature's own and he left us the beauty spots much admired by our much travelled Premier [R. J. Seddon].[55]

A less sanguine assessment would be that it was Smith's training and experience as a gardener that set him apart from self-taught and amateur home gardeners and thereby gave him the technical skills and knowledge required to transform the Domain. In her writing on early twentieth-century New Zealand gardens, Raine noted that the garden 'novelties of one generation become the conventions of the next' and fashions can change, which may explain why the Domain is not currently regarded as highly as it was once.[56] While Smith features in the *Dictionary of New Zealand Biography*, he does not in the entry on public gardens in *Te Ara* (the official New Zealand online encyclopaedia).[57] The Ashburton Domain under Smith evinced three of Strongman's five subtypes of gardens of this period: the native garden (on the island); flowering trees and shrub borders; and water gardens. However, there was no rockery or subtropical garden, although a fernery was constructed by Smith's successor.

On a return visit to Ashburton late in 1904 Smith now proffered the view that:

> New Zealand, however, possessed every class and every natural phase of scenery from the wide verdant plains and picturesque river valleys like the Wanganui to the magnificent kauri forests of the north. In addition to the towering peaks of Mount Egmont and Ruapehu and the wondrous region of Rotorua there were in the North Island, especially the numerous medieval Maori castles or pas, the history of which must be to the ethnologist of the future a subject of surpassing interest.[58]

This direct quote offers a rare insight into his thinking and indicates that he was now able to see scenic beauty in the 'natural' landscapes of New Zealand more generally. An appreciation that was manifest in embryonic form through his planting of indigenous trees and plants in the Ashburton Domain, and was undergirded by his interests in natural history and in Maori artefacts, had become more fully developed by late 1904. Smith's time in Ashburton formed an important early step towards seeing scenic beauty in the indigenous flora that was later expressed more fully in his unpublished reports for the Scenery Preservation Commission.[59]

The Scenery Preservation Commission and Pukekura Park

Although his salary had been increased to £125 early in 1901, Smith remained open to further advancement.[60] In 1902, he unsuccessfully applied for the position of Superintendent of the Dunedin City Botanical Garden and City Reserves, which went instead to David Tannock.[61] Another opportunity soon came when, after initially declining the offer, he accepted a position on the Scenery Preservation Commission (1904–1906).

The Scenery Preservation Commission was intended by Premier Seddon to, once and for all, identify and preserve the remaining scenic and historic areas of New Zealand as scenic reserves under the Scenery Preservation Act, 1903.[62] Smith was appointed secretary and, as such, as one of six commissioners, was effectively its sole full-time member. His duties involved a considerable amount of travel throughout New Zealand. Through this experience and in writing up recommendations on areas to protect, Smith's own sense of the scenic attributes of New Zealand landscapes emerged more explicitly.

After the work of the Commission was terminated by an impatient Seddon, Smith declined the earlier option of employment in the Forestry Branch of the Lands Department and instead returned to work as a gardener, finding short-term employment at Farnham House, with William and Lucy Barton at Featherston, in the Wairarapa, before taking up the position of caretaker of the Esplanade in Palmerston North, where he expressed interest in the possibilities of restoring the remnant forest alongside the Manawatu River. He was only employed there for a matter of weeks, his departure being precipitated by the mayor deriding him as a 'common cabbage gardener' at a Borough Council meeting.[63] Percy Smith, the former chair of the Scenery Preservation Commission, had in any case been endeavouring to persuade Smith to apply for the vacant position of curator of Pukekura Park in New Plymouth, where he resided in retirement. Smith secured this appointment in 1908, remaining there until his retirement in 1920. Here, indigenous plants featured much more prominently in Smith's efforts than at Ashburton. In developing the park, to some extent he was transforming it into a wilderness through large naturalistic plantings of indigenous trees. The contrasting trajectory of Smith's work at Pukekura Park has been separately addressed elsewhere, but it is useful to mention it here in order to place his earlier years at the Ashburton Domain into a broader context.[64]

Discussion

Smith's time and duties at the Ashburton Domain may be encapsulated under a number of headings, such as 'stocking the domain', 'improving the landscape' and 'policing the domain'. The Domain was still very much an open space when Smith was appointed, that not long before was deemed to be a tussocky wilderness on the edge of a small country town.

Improving the landscape brought another element of Smith's training to bear. To enhance the look of the Domain landscape early in his tenure, he was prepared to fell and heavily prune trees that had been planted 30 years before. Another of his important early tasks was to plant it up with a range of trees, shrubs, plants, and flowers. Later attention was turned to improving the paths through the domain and eventually to replanting the islands. His efforts were not, however, restricted to larger trees and shrubs. Some of the garden features that Smith laid out, for instance the rose garden near the 1897 Jubilee Gates, remain today. Stocking also

extended beyond flora to include ducks, swans and even fish. Policing the Domain was a third major task of the curator though less central to the current discussion.

It is possible to appreciate some of Smith's efforts in landscaping the Domain through the record of contemporary photographers, especially in the form of postcards, which were much in vogue in the first decade of the twentieth century. There are two caveats to their use, first, that some of the main elements of the design of the Domain were in place before Smith arrived and, second, that the act of photography itself creates a scene that was not entirely of Smith's making. Some of the postcards do, however, show features that reinforce Temple's assessment from Burghley of Smith's skill in geometric layout and the creation of naturalistic and picturesque landscapes.

A final, but not insignificant, point relates to Smith's attitude to the indigenous flora. His passionate interest in natural history signals a growing familiarity with the indigenous flora and fauna of New Zealand. However, the lengthy list of donations of seeds and plants in the Domain Board minutes was comprised predominantly of familiar British and European species, with the exception of 'native [i.e. Australian] shrubs' from Brisbane in 1895. Only six gifts were of indigenous species, including 'young cabbage palms' (1896) and 'alpine plants' from Mr Bates (1896), ferns from Mr King in Springburn (1897), unspecified plants from Acland at Mt Peel (where Smith had previously been gardener), seeds from Mr Williams the Crown Lands Ranger in Timaru (1898), and further gifts from Mt Peel (1902). The appreciation of indigenous plants in the district was still comparatively rare.

In 1903, Smith was involved in a project that explicitly involved indigenous trees and took the form of trimming back the undergrowth on the large island, which included willows, and in their place planting native trees, which the local paper claimed would 'greatly improve the appearance of the island' (figure 5).[65] This constituted Raine's 'native garden', but in Ashburton Domain it was for most visitors only one of a number of garden types and had no special purchase on their aesthetic sensibilities. Those who lauded Smith's ability to produce scenic landscapes in the Domain make no specific mention of indigenous species. This may reflect a view that if the tree fitted, then it was irrelevant where it came from. Newspaper endorsement of Smith's work on the island in 1903, however, suggests attitudes were somewhat in flux, consistent with environmental historian Paul Star's identification of the use of indigenous flora and fauna as symbols of colonial difference.[66]

FIGURE 5. *The 'big island' in the main pond at Ashburton Domain. Smith commenced the planting of this with indigenous trees to form a 'native garden' in 1903 and it has remained something of a refuge for them in the domain. Source: photograph by author.*

Period photographs of the Domain show that various other indigenous plants were evident, including cabbage trees, flax, and toetoe.[67] Within the constraints of restricted funding and a limited choice of species, as well as the problems of growing trees on what was still a fairly wind-exposed site, not to mention the close control that the Domain Board tended to exert over Smith's day-to-day activities, there are signs of a growing regard for the indigenous flora and for New Zealand landscapes. Ashburton Domain is an early step in the development of Smith's ideas about scenic beauty in the indigenous flora, which he later expressed more fully in his reports for the Scenery Preservation Commission.[68] But even these were not the final stage for subsequently, in Palmerston North and New Plymouth, he sought to recreate sizable naturalistic areas of indigenous forest alongside formal ornamental beds of exotic shrubs and flowers.

Previously I have presented Ashburton Domain and Pukekura Park in New Plymouth as two strongly contrasting public domains, suggesting that in Ashburton, Smith was still operating entirely according to his British training, but made greater use of indigenous trees and shrubs at Pukekura Park and that this mirrored a change in his appreciation of what constituted scenic beauty in the New Zealand landscape.[69] The more detailed treatment of Ashburton Domain undertaken here suggests a qualification is necessary, to the effect that Smith did begin to experiment with indigenous trees at Ashburton, albeit initially in a limited way through the 'native garden'. His capacity to do more at that stage was constrained on multiple fronts: suitable plants were in short-supply, the climate was more restricting than that he would encounter in New Plymouth, and the Domain board maintained a close oversight over his day-to-day work routines. For all that, understanding more clearly what Smith accomplished in the Ashburton Domain positions his efforts on the Scenery Preservation Commission and his later curatorship of Pukekura Park.

Conclusion

For residents of Ashburton, the act of planting trees in itself would have constituted a significant improvement to the Domain, given the exposed position of the town site on the flat and initially tussock-covered Canterbury plains. On more than one occasion the newly formed cricket pitch surface was blown away. That said, Smith brought a new level of expertise and industry to his time as curator, both in improving the layout of the Domain and in establishing formal flower beds, but also in planting trees and labelling specimens for educational purposes. He put an enduring mark on a Domain that was much admired not only locally but further afield and contributed to his appointment to the Scenery Preservation Commission and subsequent employment at Pukekura Park. A hallmark of his work was the creation of a number of scenic vistas and picturesque landscapes within the Domain. This transformation of the Domain landscape was important and locally significant, and testament to his training and competence, but arguably in the long term it was not the most telling outcome of Smith's time and efforts as curator. Instead, this lies in Smith's increasing engagement with the indigenous flora and fauna, something that gathered momentum during his time in Ashburton, but found fullest expression when he served as commissioner and at Pukekura Park.

Acknowledgements

The assistance of the staff of the Ashburton District Council Archives, the Ashburton Museum, and the Ashburton Public Library is gratefully acknowledged. My thanks also go to Dr Alistair Clement for preparing the map.

Funding

This work was supported by a Massey University Research Fund Grant, which assisted with travel and other research costs.

Disclosure statement

No potential conflict of interest was reported by the author.

NOTES

1. Dawn McLeod, *The Gardener's Scotland* (Edinburgh: Blackwood, 1977).
2. William Temple, Testimonial for W. W. Smith, 9 December 1874. Correspondence 1870-Dec 1920 ARC 1-163/2 Smith W.W. M4/1/2, Smith Papers Puke Ariki, New Plymouth.
3. E. C. Till, 'The Development of the Park and Gardens at Burghley', *Garden History*, xix/2, 1991, pp. 128–145; p. 141.
4. Roger Turner, *Capability Brown and the Eighteenth-Century English Landscape* (London: Weidenfeld and Nicholson, 1985).
5. Thelma Strongman, *The Gardens of Canterbury* (Wellington: Reed, 1984); Katherine Raine, '1860s–1900 Victorian Gardens', in Matthew Bradbury (ed.), *A History of the Garden in New Zealand* (Auckland: Viking, 1995), pp. 87–107; and Katherine Raine, '1900–1920 Early Twentieth Century Gardens', in Bradbury (ed.), *A History of the Garden in New Zealand*, pp. 113–134.
6. Raine, '1860s–1900 Victorian Gardens', p. 98.
7. Raine, '1860s–1900 Victorian Gardens', p. 101.
8. Strongman, *The Gardens of Canterbury*, p. 143. On the six categories, see pp. 143–146.
9. See for example, Winsome Shepherd and Walter Cook, *The Botanic Garden, Wellington: A New Zealand History 1840–1987* (Wellington: Millwood Press, 1988), or the older M. J. Barnett, H. G. Gilpin and L. J. Metcalf (eds), *A Garden Century: The Christchurch Botanic Gardens, 1863–1963* (Christchurch: Christchurch City Council, 1963), and A. B. Scanlan, *Pukekura: A Centennial History of Pukekura Park and Brooklands* (New Plymouth: New Plymouth City Council, 1979). A more expansive overview is provided by Paul Tritenbach, *Botanic Gardens and Parks in New Zealand: An Illustrated Record* (Auckland: Excellence Press, 1987). Public parks are succinctly discussed by John Adam, 'Parks and Public Gardens 1840s to 1860s', in Bradbury (ed.), *A History of the Garden in New Zealand*, pp. 84–85, and John Adam, 'Parks and Public Gardens 1860s to 1900', in Bradbury (ed.), *A History of the Garden in New Zealand*, pp. 110–111. See also, Franklin Ginn, 'Colonial Transformations: Nature, Progress and Science in the Christchurch Botanic Gardens', *New Zealand Geographer*, liv, 2009, pp. 35–47; while private gardens are a related subject, see Helen M. Leach, 'Exotic Natives and Contrived Wild Gardens: The Twentieth-Century Home Garden', in Tom Brooking and Eric Pawson (eds), *Environmental Histories of New Zealand* (Melbourne: Oxford University Press, 2002), pp. 214–229.
10. For example, Rupert Tipples, *Colonial Landscape Gardener: Alfred Buxton of Christchurch, New Zealand, 1872–1950* (Lincoln: Lincoln College, 1989); Alison Evans, 'Tannock, David', in the *Dictionary of New Zealand Biography* http://www.TeAra.govt.nz/en/biographies/3t3/tannock-david; and Allen Hale, *Pioneer Nurserymen of New Zealand: Compiled for the 50th Anniversary of the New Zealand Horticultural Trades Association (Inc.)* (Wellington: A. H. and A. W. Reed, 1955).
11. Strongman, *The Gardens of Canterbury*, p. 81.
12. Strongman, *The Gardens of Canterbury*, p. 90.
13. Strongman, *The Gardens of Canterbury*, p. 90.
14. Personal observation, 2011. Sad to say, my packet did not sprout.
15. Oliver Gillespie, *South Canterbury A Record of Settlement* (Timaru: South Canterbury Centennial Project, 1971), p. 122.
16. Johannes C. Anderson, *A Jubilee History of South Canterbury* (Christchurch: Whitcombe and Tombs, 1916), p. 48; and W. W. Smith, 'On the Birds of Lake Brunner District', *Transactions of the New Zealand Institute*, 21, 1888, pp. 205–224.
17. 'The Horticultural Society's Autumn Show', *North Otago Times* (18 March 1882), p. 2.
18. W. W. Smith, 'On the Birds of Lake Brunner District', pp. 205–224.
19. W. W. Smith, 'Notes on *Eristalis tenax* in New Zealand', *Entomologists' Monthly Magazine*, xxvi, 1890, pp. 240–242.
20. Based on an analysis of newspaper clippings in Smith's papers in Puke Ariki, from incidental details in his natural history publications and from a search of Otago and Canterbury newspapers on Paperspast.

21. Ashburton Domain Board Minutes for 8 June 1894, Ashburton Museum Archives, Ashburton.

22. Ross Galbreath, *Walter Buller: The Reluctant Conservationist* (Wellington: GP Books, 1989).

23. Smith, 'On *Hiperacidea novae-zealandae* and *H. brunnera*', *Transactions of the New Zealand Institute*, 16, 1883, pp. 318–322; and Smith, 'On the Birds of Lake Brunner District'.

24. 'Otago Museum Requirements', *Otago Witness* (4 April 1885), p. 17, and 'The Museum', *Otago Witness* (13 March 1886), p. 8.

25. Smith, 'On the Birds of Lake Brunner District'; Smith, 'Notes on Certain Species of New-Zealand birds', *Ibis*, v/20 (sixth series), 1893, pp. 509–521; and Smith, 'Notes on *Eristalis tenax*'.

26. Smith, 'The Land of the Moa', *North Otago Times* (28 January 1892), p. 3.

27. Kerry Howe, *The Quest for Origins* (Auckland: Penguin, 2008), offers a critical reassessment of various theories.

28. Smith, 'Origins of the Canterbury Rock Drawings [Note]', *Journal of the Polynesian Society*, vi, 1897, pp. 158–159; and Smith, 'On Ancient Maori Relicts from Canterbury, New Zealand', *Transactions of the New Zealand Institute*, 33, 1900, pp. 426–433.

29. Galbreath, *Walter Buller*, p. 211.

30. W. W. Smith to S. Percy Smith, 11 August 1903. LS1 51887/84 Straight Number Series, Wellington: Archives New Zealand.

31. David Tannock, *Manual of New Zealand Gardening* (Auckland: Whitcombe and Tombs, 1914).

32. *Cyclopaedia of New Zealand* (Christchurch: Cyclopaedia of New Zealand Company, 1903), p. 815.

33. 'The Ashburton Domain', *Ashburton Guardian* (28 March 1904), p. 2.

34. Other important figures to 1950 included James Young, who although curator only from 1904 to 1907 greatly expanded the walkway system before taking up the position of Curator of the Christchurch Botanical Gardens, and Dennis Leigh (1938–1949), who had experience at Kew and later became curator at Nelson. Leigh expanded the number and variety of flowers including the planting of a vast number of daffodils. *Ashburton Domain and Gardens 150 Years* (Ashburton: Ashburton District Council, 2014), p. 6.

35. Ashburton Domain Board Minutes for 11 February 1895.

36. 'The Ashburton Domain', *Ashburton Guardian* (28 March 1904), p. 2.

37. W. Harry Scotter, *Ashburton, A History with records of Town and County* (Ashburton: Ashburton Borough and County Councils, 1972), p. 128.

38. 'The Ashburton Domain', *Ashburton Guardian* (28 March 1904), p. 2.

39. H. Sealy, reference for Dunedin job application, 2 September 1902. Correspondence 1870-Dec 1920 ARC 1-163/2 Smith W.W. M4/1/2 Smith Papers Puke Ariki, New Plymouth.

40. 'Borough Council', *Ashburton Guardian* (10 September 1895), p. 2; and Ashburton Domain Board Minutes from 1886 to 1906.

41. 'The Ashburton Domain', *Ashburton Guardian* (28 March 1904), p. 2.

42. Ashburton Domain Board Minutes for 20 May 1895, 20 February 1896, 8 January 1900, 6 August 1900 and 25 June 1900.

43. 'Acclimatisation Society', *Ashburton Guardian* (1 October 1903), p. 2; and *The Cyclopaedia of New Zealand*, p. 815.

44. 'Local and General', *Ashburton Guardian* (28 May 1903), p. 2.

45. 'Local and General', *Ashburton Guardian* (26 June 1903), p. 2.

46. Smith, 'Plants Naturalised in the County of Ashburton', *Transactions of the New Zealand Institute*, 36, 1903, pp. 203–225.

47. 'A Notable Pleasure Resort', *Ashburton Guardian* (26 February 1903), p. 2.

48. Starling Gray, 'Our Domain', *Ashburton Guardian* (3 October 1896), p. 3.

49. Starling Gray, 'Our Domain', *Ashburton Guardian* (3 October 1896), p. 3.

50. Starling Gray, 'Our Domain', *Ashburton Guardian* (3 October 1896), p. 3.

51. 'A Notable Pleasure Resort', *Ashburton Guardian* (26 February 1903), p. 2.

52. 'The Domain Curator', *Ashburton Guardian* (12 March 1904), p. 2.

53. 'The Domain Curator', *Ashburton Guardian* (12 March 1904), p. 2.

54. 'The Domain Curator', *Ashburton Guardian* (12 March 1904), p. 2.

55. 'Rate Payers and the Domain', *Ashburton Guardian* (15 July 1905), p. 2.

56. Raine, '1900–1920 Early Twentieth Century Gardens', p. 98; Maggie Wassilieff, 'Public Gardens', in *Te Ara — the Encyclopedia of New Zealand*, http://www.TeAra.govt.nz/en/public-gardens.

57. Ross Galbreath, 'Smith, William Walter, 1852–1942', in the *Dictionary of New Zealand Biography: Volume 3, 1901–1920* (Wellington: Department of Internal Affairs, 1996), p. 482.

58. 'Scenery Preservation', *Ashburton Guardian* (11 October 1904), p. 2.

59. Michael Roche, 'Seeing Scenic New Zealand: W. W. Smith and the Scenery Preservation Commission 1904–1906', Paper presented to the XV International Conference of Historical Geographers, Charles University, Prague, 6–10 August 2012.

60. Ashburton Domain Board Minutes, 4 February 1901.

61. Charles Chilton to J. Park (Mayor of Dunedin), 15 November 1902. Correspondence 1870 - Dec 1920 ARC 1-163/2 Smith W.W. M4/1/2 Smith Papers Puke Ariki, New Plymouth.

62. Tony Nightingale and Paul Dingwall, *Our Picturesque Heritage: 100 Years of Scenery Preservation in New Zealand* (Wellington: Science & Research Unit, Department of Conservation, 2003). Premier Seddon's motives are further discussed in Tom Brooking, *Richard Seddon: King of God's Own the Life and Times of New Zealand's Longest Serving Prime Minister* (Auckland: Penguin, 2014).

63. 'That Common Cabbage Planter', *Manawatu Times* (11 March 1908), p. 5.

64. Michael Roche, 'Transforming the Colonial Settlement with Parks & Domains: Scenic Beauty in Two New Zealand Towns 1894 to 1920', in A. Gaynor, E. Gralton, J. Gregory and S. McQuade (eds), *Proceedings of the 11ᵗʰ Urban History/Planning History Conference* (Perth: State Library of Western Australia, University of Western Australia, 2012), pp. 293–305.

65. 'Borough Council', *Ashburton Guardian* (28 May 1903), p. 2.

66. Paul Star, 'Native Bird Protection, National Identity and the Rise of Preservation in New Zealand to 1914', *New Zealand Journal of History*, xxxvi/2, 2002, pp. 123–133; and Paul Star, 'Native Forest and the Rise of Preservation in New Zealand (1903–1913)', *Environment and History*, viii/3, 2002, pp. 275–294.

67. New Zealand flax (*Phormium tenax*) is unrelated to the European linen flax (*Linum usitatissimum*).

68. Roche, 'Seeing Scenic New Zealand'.

69. Roche, 'Transforming the Colonial Settlement'.

The cultural history of the garden gnome in New Zealand

IAN C. DUGGAN

Introduction

Garden gnomes are a dominant, and iconic, component of the suburban garden statuary in many parts of the world, including England, North America, Australia and New Zealand. These ornaments have traditionally been depicted as bearded 'dwarf-like' human figures, male, with a red pointed hat, although this representation has diversified in recent years.[1] Garden gnomes have long been ignored by garden historians, and have only recently been gaining academic attention, primarily in England.[2] However, little is known of their usage and popularity elsewhere. This article will address this lacuna by examining their introduction and cultural significance in New Zealand. I identify two key periods in their New Zealand usage: (1) prior to the 1940s, when garden gnomes were expensive and utilized by a wealthy elite; and (2) since the 1950s, when they became more affordable and popular in suburban gardens. The article outlines key differences in the history and role of garden gnomes in New Zealand and England.

In her book *Garden Gnomes: A History*,[3] Twigs Way provided details of the early origins and development of the garden gnome. The first gnome-like garden figures date back to at least the seventeenth century, although these differed from modern gnomes in both their attire and activities. Although their origins are diverse, with similar statuettes known from Denmark and Sweden, the modern garden gnome owes much of its appearance to Germany. The first ornaments we might recognize as 'garden gnomes' were being made in Germany by the late eighteenth century, although these were not intended at the time as garden ornaments, but as house dwarfs — that is, for indoor display — and were made of porcelain. The first purpose-built gnomes constructed for actual outdoor use date to around 1840, from Germany or Poland. These were encountered by English tourists, and exported across Europe and to America by 1860. However, although not constructed with the intention of outdoor usage, house dwarfs became the first gnomes in English gardens. Credit for the first outdoor garden gnomes in England goes to Sir Charles Isham, an eccentric, non-smoking vegetarian — and hater of all blood-sports — who inherited 'Lamport Hall' in Northamptonshire following the suicide of his brother in 1846. In 1847, he initiated a rockery in which, to complement its dwarf and alpine plants, he introduced a number of house dwarfs, particularly to the caverns and tunnels. This garden featured in magazines in the late 1800s, and these articles are thought to be responsible for the popularity of garden gnomes in English stately homes from this time. Isham died in 1903, at which time his gnomes were banished. Isham's daughter was reputed to have hated the gnomes and had them destroyed, but missed one. This gnome, affectionately known as 'Lampy', was well hidden in a crevice in the garden, and was not rediscovered until after the Second World War, during restoration of the garden by Sir Gyles Isham. 'Lampy' is now considered to be England's oldest gnome. Garden gnomes fell from fashion in England following the First World War due to their association with Germany; German soldiers used the gnomes as mascots, and were photographed with them as good luck charms.

Thus, only a small market existed for garden gnomes in the 1920s, but by the 1930s the popularity of gnomes in England again increased, although now primarily in suburban settings.[4]

The earliest garden gnomes in New Zealand

Garden gnomes owe their existence in New Zealand to British colonization, and are still freely available for sale in garden stores and via the internet. Despite their great popularity from the 1930s, no mention is made of gnomes in most of New Zealand's garden histories.[5] This is in common with England where, as Twigs Way notes, at the same time as they were 'overpopulating the suburbs', garden gnomes ' … were being selectively written out of garden history'.[6] Gnomes in New Zealand gardens, although still commonly seen today, appear to have similarly been ignored by the country's garden historians. This is in contrast to a burgeoning literature on public statuary, which documents the good and the great who have graced New Zealand's shores, as well as the sculptors who made them.[7] Within New Zealand gardens, water features, and structures, such as bridges, pergolas and arches, are commonly described,[8] while less ubiquitous garden ornaments, such as concrete Japanese lanterns[9] and Chinese 'dogs of Fu' have also gained attention.[10] Snobbishness among New Zealand garden historians may partially account for their almost total silence on garden gnomes. It is time, then, to rescue garden gnomes from the condescension of the present and to restore them to their rightful place in histories of the garden in New Zealand.

Through its colonization and the subsequent large-scale immigration of British, New Zealand has strong connections with the UK, and especially England, which provided the place of origin of most settlers.[11] Given this, and the exclusion in both countries of gnomes from garden history texts, the trends in popularity and status of gnomes might also be expected to parallel one another. However, this is apparently not the case. The appearance and availability of gnomes in New Zealand coincided with their English resurgence in the early 1930s, and not with the latter half of the 1800s when they were first popularized in England. Up until 1840, when the first purpose-built gnomes were being produced in Europe, New Zealand was inhabited by only a small number of non-*Māori* (or *Pākehā*). The population of Pākehā grew rapidly from 1849, however, from a little over 2000 to 500 000 by 1881. Foreign-born outnumbered colonial-born in the 1881 census, with most being of English, Scottish and Irish extraction. Those from England were primarily from London, and the southeast and southwest,[12] while the 'hotbed' of gnomes in England at the time was around Northamptonshire.[13] Along with social and geographical factors, because labourers and tradesmen predominated over the gentry among New Zealand immigrants, garden gnomes were unlikely to have been imported during their first wave of English popularity. Even a New Zealand book on rockery gardens produced in 1923, where garden gnomes might have been most probably encountered, fails to mention them.[14]

Despite an absence of earlier records of garden gnomes, their presence was commonly noted from the early 1930s. The earliest record of garden gnomes in New Zealand is from an article in an Auckland newspaper, the *New Zealand Herald*, from early 1931. The article noted:

> A host of unique novelties for the garden or 'outdoor living room' are on display on the fifth floor of the premises of Messrs John Court, Limited, and are attracting considerable attention. The ornaments, which are in the form of statuettes of gnomes, animals and birds, are made of gutta-percha and are gaily embellished with special rich coloured paints that will withstand all conditions of weather and the heat of the summer sun.

> Artistically arranged amid beautiful pot plants and set on an artificial grass lawn the figures present a most appealing spectacle. Perhaps the most realistic ornament is that which shows a merry little gnome fishing in a crystal-clear goldfish bowl. Other appropriate garden furnishings, all fine specimens of the modern home decorator's art, include figures of gnomes busily engaging in various garden pursuits; ever-popular rabbits and deer, a prancing fox terrier puppy, a hen with her brood of fluffy chicks, and numerous birds.[15]

This article is likely incorrect on the material the gnomes were made of, confusing the terms gutta-percha, a form of natural rubber or latex, with terracotta, the clay used for pottery, which was in widespread usage in gnome construction at this time.[16] This error itself is indicative of how uncommon garden gnomes were in New Zealand. Further, an advertisement for the same company later in 1931 featured three gnomes, each with distinctive hats and wheelbarrows, consistent with the terracotta gnomes of the German 'Heissner' company (figure 1).[17] Heissner was one of the best-known names in gnome manufacturing at the time, having begun gnome production

The *Quaintest* Garden Ornaments

Know us by the
Swastika . . . J.C.L.
Symbol of Satisfaction in Quality . . .
Value . . . Service.

Made from Genuine **TERRA COTTA.**

Summer nights will soon be here. Embellish the garden with quaint Gnomes, Rabbits, Garden Seats, in tree trunk and monster toadstool forms, etc. A superlative collection for your choosing at J.C.L.

Quaint Garden Ornaments, artistically coloured with specially prepared paints to withstand all weathers—impervious to rain, hail, or shine.

A. **Gnome With Barrow,** as illus. } 29in. high, **99/6**

B. **Gnome Carrying Basket,** as illus. **Price, 47/6**

C. **Gnome, Reclining,** as illus. 12in. long, **29/6**

D. **Rabbit,** as illus. 14in. high, **39/6**

We Carry a Comprehensive Selection of Quaint Pieces, including:
Garden Seats in form of Tree Trunks, 55/-: In form of Toad Stools; small size, 35/-; large size, 59/6 . Deer: 11/6 and 47/6 . Pigeons, 17/6 . Fox Terriers, 3/6 . Storks, from 50/- . Parrots on Perches, 45/-. Other pieces from as low as 3/6.

Art. China Department—For Garden Ornaments . Stuart Crystal . Glassware . China . Pottery . Etc.

FIGURE 1. *Advertisement for terracotta gnomes, and other figures, available from John Court Ltd, Auckland. Auckland Star, 14 November 1931. Source: National Library of New Zealand.*

in 1872, and is still extant today.[18] The seller of the Heissner gnomes in 1931 New Zealand, John Court Limited, was one of the leading firms in Auckland, occupying an eight-storey building in a prominent position on a corner of Queen Street in the central business district.[19] Despite the company having a tradition of good service, quality merchandise and reasonable prices,[20] the cost of the garden gnomes was likely unaffordable to many at this time, ranging from 29/6 (or £1 9s 6d) for a modest 12-inch gnome, to 99/6 (or £4 19s 6d) for a 29-inch gnome (current decimal equivalents £1.47 and £4.97,

respectively).[21] What is more, the effects of the so-called Great Depression were beginning to take effect in New Zealand, meaning such luxuries would not have been considered appropriate, or affordable, purchases.[22] The seemingly democratic origins of the modern garden gnome in New Zealand, then, began with an expensive figure produced for an elite market.

Garden gnomes were certainly not the maligned figures of shame in the 1930s that they — arguably — later became. In November 1932 Wellington's *Evening Post* reported:

Great pleasure was given yesterday afternoon to many Wellington women by a visit to Miss Black's new floral studio in Woodward Street. Miss Black having shown her claim to be an advisor in house furnishings and decorations is a sound one. A simulated grass plot contained tree effects with birds, quaint gnomes, a dainty blue pigeon, frogs, rabbits, hares, and squirrels, all ready for garden decoration. A central adornment was a stand of goldfish, and on a huge mushroom was a frog and attendant gnome.[23]

Garden gnomes for the wealthy

New Zealand was a wealthier nation in the 1930s than it was in the late-1800s.[24] In that period, many settlers succeeded in 'getting on', some earning a small fortune through their own efforts.[25] This applied to several early gnome owners. The first garden gnomes in New Zealand were bought by elites, and became high-status objects mentioned in newspaper reports of the private homes of prominent, wealthy individuals, who would open their homes for charity and community events. The earliest mention of garden gnomes, *in situ*, is at Wellington's 'Homewood' gardens, owned by Mr Benjamin Sutherland, a New Zealand-born chain-store pioneer.[26] Sutherland bought 'Homewood', in the Wellington suburb of Karori, in 1928, commissioning Christchurch landscape gardener Alfred Buxton to 'lay out the whole garden, and no expense was spared'.[27] In the early 1930s, between 12 and 20 men worked on the garden for a period of two years to build, among other structures, three walled gardens, as well as grottos and glow-worm caves, a large white-tiled swimming pool and 18 aviaries housing hundreds of birds.[28] With a bridge and pergola located next to a pond filled with water lilies, much of his gardens had *Japonisme* overtones.[29]

Sutherland, a philanthropist, held regular open days and charity events at his residence. In March 1932, the *Evening Post* correspondent described a fundraising event for the 'Free Ambulance', and a special garden inhabitant: 'The grey, battlemented house, with the stone lions in front, was well set in vivid borders of flowers, while the sardonic looking gnome, who presides over the fish pond, seemed to smile at the rain … .'[30] Gardens gnomes were in place for many years at 'Homewood', and there must have been a number of them on the property. In March 1941, a report on the 'Homewood' grounds, opened in aid of the 'Karori Soldiers Welfare Association', again describes the resident gnomes. The article listed Homewood's attractions, including its begonias, beautiful foreign birds, and glow-worm cave, and advised visitors 'do not miss the gnomes'.[31] It is unclear how many gnomes were in place at 'Homewood'. However, the *Evening Post* noted in 1941 that the grotto fernery was considered 'particularly outstanding', and 'the glow-worm cave, with its groups of gnomes watching water displays, is another outstanding attraction'.[32] Another contemporary description, from a programme for an open day in aid of the Free Kindergarten in November 1941, recorded 'a gnome village'. It continues:

> Taking a path to the right, and crossing a rustic bridge, one comes to the Glow-worm Cave, cunningly fashioned in the hillside. At the entrance is a most charming scene. Disposed in care-free attitudes around fine tiny fountains are groups several gnomes, watching with delighted attention balls kept in play by jets of water on which are played vari-coloured lights.[33]

Gnomes were therefore found in at least two locations at Homewood.

As noted, Sutherland was a wealthy, self-made man. His Self Help Cooperative Limited was a successful New Zealand grocery stores chain. Founded in Wellington in 1922, it expanded from seven Self Help stores in Wellington within one year of opening to 56 throughout New Zealand by 1929, and 100 by 1931.[34] The particular attention given to gnomes in accounts of Sutherland's garden, as well as their prominent display, suggests that not only were they uncommon in this period but also that they were symbols of elite status.

This is highlighted because all other reports from the 1930s described gnomes found in the gardens of the wealthy. In September 1932, *New Zealand Herald* ran a story about: 'One of the most beautiful Gardens in New Zealand', at the home of Frank Crossley Mappin, in the Auckland suburb of Epsom.[35]

After receiving an unexpected, and substantial, inheritance from an uncle in 1920, the former fruit grower bought a six-acre property with his wife, Ruby, replacing an unsuitable house with a larger one, which they named 'Birchlands'. Adding an adjoining six acres, they made a twelve-acre estate, which the Mappins turned into a private park that included several glasshouses and which required three gardeners to keep up. Within this garden, the *Herald* reported of 'The Bluebell Dell': 'Following the same path you pass rock walls, hung with scented herbs, thyme, sage and catmint, and in a moment you are standing at the edge of a little pond shaded by tall trees, bordered by primroses. A tiny gnome is fishing there, a stone stork peers with hopeful eye into dim green waters, each exhibiting the immemorial patience and optimism of the born angler.'[36] Again, it seems likely that the rarity of gnomes made them attractive to wealthy garden owners, keen to show off to others their possession of the latest expensive garden ornaments. Mappin was also a collector of Chinese and Japanese art,[37] and for garden gnomes to be associated with such a prominent and serious collector suggests that the gnomes, too, were perceived as elite, high-status objects.

As small, easily moved and expensive objects, garden gnomes proved irresistible to thieves and pranksters. A 'Lost and Found' notice in the Wellington newspaper *Evening Post* in 1935, offered a 'good reward' for 'information leading to the recovery of Two Garden Ornaments, one Gnome and Toadstool, removed from the Garden of 186 The Terrace'. This advertisement provides further evidence that gnomes were primarily for the wealthy in New Zealand.[38] At this time, the property, 'Fern Glen', was owned by a medical doctor, William E. Herbert.[39] The advertisement also provides the first documentation of theft, which has formed an enduring theme in the history of garden gnomes in New Zealand.

On 13 December 1939, the *Hutt News* reported another theft, advertising a reward of £5 for a 4 ft high garden gnome from a private residence at 29 Old Military Rd, Lower Hutt.[40] Three days later, the reward was again offered in the *Evening Post*, which included a photo of the missing gnome (figure 2). Seated in a relaxed fashion on a tree stump, its sharply pointed upright hat, pointed shoes and rough beard, suggest the gnome is again the product of another German company, on this occasion Eckardt & Mentz, which manufactured gnomes from at least the 1860s through to 1945.[41] Like Herbert's gnome, it would have been constructed of terracotta. The owner who sought its return was David Livingstone Harrower,

STOLEN!

£5 Reward

The above garden ornament, standing 4 feet high, was stolen from the residence, 29 Old Military Road, Lower Hutt, Wellington. A reward of £5 will be paid for information leading to its recovery. Please inform the nearest Police Station.

FIGURE 2. *The stolen gnome and reward offer of David Livingstone Harrower of 'Eversleigh', Lower Hutt. Evening Post, 16 December 1939. Source: National Library of New Zealand.*

the partner-manager of 'The Don Tailors' in Wellington, who resided at 'Eversleigh'. With a generous reward offered, the Honourable Secretary of the 'Lower Hutt Liedertafel',[42] who made at least one business trip to England to secure fabric,[43] was clearly — as with the previous gnome owners — moderately wealthy. The loss of both Herbert's and Livingstones' gnomes shows that theft in New Zealand is not simply a modern phenomenon, but has existed from the first availability of the garden gnome.[44]

Another account of the tenth annual meeting of the Alpine and Rock Garden Society in 1937 describes the garden of Mr and Mrs Hope Barnes Gibbons. Before the formal business began, the guests were treated to a tour of the Gibbons' gardens at their home 'Ngaroma', in the Wellington suburb of Lyall Bay. The tour took in a rock garden in his 'smaller lounge': 'At one side was a quaint gnome, who smoked a pipe and looked on at the ducks, rabbits and other animals who peeped out from various places in the garden.'[45] Again, like Sutherland, Mappin, Herbert and Livingstone, the Barnes Gibbons were a wealthy, up-and-coming family. 'Hope' was one of four sons of the successful brewer, businessman, and mayor of Wanganui, Hopeful Gibbons. The family worked together in their commercial undertakings, acquiring a series of businesses over time, including owning a controlling interest in Wellington's Colonial Motor Company by 1917, the holder of the Ford franchise for New Zealand. By 1920 they had constructed the first purpose-built car assembly plant in New Zealand.[46]

Finally, an auction at the Lower Hutt residence 'Lansdowne', of Mr William J. Mason, in January 1940, provides further evidence of garden gnomes as elite items.[47] Included in his belongings were garden ornaments, including '3 dwarfs', along with two rabbits, a squirrel and a blackbird. Their grounds, also, were used for charity events in the late 1930s.[48] The auction notice for the property, in late 1939, advertised it as one of the finest residential properties in Lower Hutt, with its grounds laid out by Alfred Buxton.[49] Buxton, who had also designed 'Homewood', was the most significant landscape gardener in New Zealand in the first half of the twentieth century, and a Buxton landscape was in itself a symbol of that affluence.[50] His large-scale designs, with gardens joined by walks and ornamental steps, included planted trees in clumps and belts, and aquatic features such as ponds, fountains and waterfalls,[51] while there was also a strong influence of *Japonisme* in many of his designs and garden statues.[52] These designs allowed for the introduction of such things as ornamental bridges,[53] and seemingly also ornaments of other types. The excesses of Buxton's landscaped gardens contrasted with the typical depression-era quarter-acre garden of New Zealand's suburban dwellers.[54] The use of rare and expensive garden gnomes from Germany was most likely — as with the aspects of orientalism in both Buxton's gardens and Mappin's art collection — a further statement of the garden owner's wealth and elite standing.

Overall, these first accounts of garden gnomes in New Zealand suggest that they were owned by successful and wealthy people; philanthropists and socialites who opened their doors to the public for social events and who therefore conspicuously displayed their wealth to the public. This was markedly different from England in the 1930s. By this decade, garden gnomes had lost some of their privileged status, having moved from stately homes onto suburban lawns and into amusement parks.[55] Although the gardens of New Zealand suburban homes are not likely to have had their garden statuary reported in the newspapers, which could skew our view of gnome ownership, there are a number of other lines of evidence that suggest that garden gnomes in 1930s New Zealand were primarily the property of the wealthy. For example, the prices for gnomes listed by John Court Limited would have been beyond the means of the average suburbanite, while the quote above proclaiming Miss Black a sound advisor in house furnishings and decorations — with her garden gnome as a central adornment — suggests that gnomes at this stage were associated with people of a particular standing in society. The fact that they were even mentioned in newspaper reports suggests that garden gnomes were noteworthy, and not simply an ordinary sight in gardens. And the properties in which they resided were substantial, architecturally designed homes and gardens; three of which survive today as official residences. Benjamin Sutherland's 'Homewood' property was sold to the British Government as the residence of the British High Commissioner to New Zealand in 1958, a purpose for which it continues to serve. Frank Mappin's 'Birchlands' property is now 'Government House', gifted by Mappin to Her Majesty the Queen in 1969, and which became the official Auckland residence of the Governor General.[56] It is currently used for many official functions, including the welcoming of visiting heads of state.[57] And Gibbons 'Ngaroma' property is now the Apostolic Nunciature, the embassy of the Holy See. Only Herbert's residence at the Terrace has gone, demolished in 1963, and replaced by a thirteen-storey apartment building.[58]

Many of the early reports of gnomes were made in newspapers. Unlike Britain, newspapers in New Zealand were seemingly not aimed at any particular sector of society. As such, the reports of gnomes in gardens were available to all literate New Zealanders, not just the wealthy. For example, the New Zealand working class in the nineteenth and early twentieth centuries had a relatively high standard of living, and, unlike in Britain at this time, could regularly eat meat, own their own homes, own horses and (later) cars.[59] In addition, the New Zealand population had high literacy rates, with the National Education Act of 1877 allowing for all New Zealanders to be taught to read and write.[60]

The 1940s: wider availability, and a loss of status?

Garden gnomes increased in availability in the 1940s, probably as a result of cheaper construction methods making them more affordable. For example, 'The Rangatira', a florist and pet store in Lower Cuba Street, Wellington, ran a series of advertisements in May 1940 on its comprehensive stock of gnomes and garden ornaments, designed to 'add charm to garden or sun porch'.[61] From June to September 1940, advertisements began to appear in *New Zealand Herald's* 'Garden Needs' section for 'Gnomes, bird baths, animals, [and] garden ornaments', from Centennial Florist at the bus station.[62] In 1945, Smith & Smith Limited were also advertising garden gnomes among their garden ornaments in a newspaper advertisement.[63] Garden catalogues, too, began to carry advertisements later in this decade. The Griffiths gardening catalogue for Spring 1946 advertised gnomes and other ornaments as Christmas gift suggestions.[64] Three different gnomes were available from the company at this time, of between 11 and 12 inches in height, for 25s (or £1 5s; now £1.25) each, including postage. Although 'strongly made and gaily coloured', and with scarlet caps, brown trousers and blue coats, it is unclear who the maker of these gnomes was, or from what they were constructed. However, the gnomes were seemingly no longer produced by the major German makers who had been predominant in the 1930s. For example, Griffiths' Winter 1947 catalogue pictured a set of five very different gnomes, and made it clear that these were 'strongly made of concrete composition'.[65] Labelled 'Fishing', 'Simon', 'Joker', 'Tired', and 'Hoppy', and measuring 14 inches, these concrete gnomes were available for 22/6 (or £1 2s 6d; now £1.12) each, or 25s, with postage paid, or for a smaller reclining gnome six by ten inches, 16s (now £0.80), or 17/6 (now £0.87), with postage. Taking inflation into account, the costs of these concrete gnomes are around half that of the terracotta gnomes sold by John Court Ltd in 1931, making them much more affordable for the suburban masses than in the past. Concrete provided advantages over terracotta, in that the gnomes were quicker to produce, allowing for a mass market; however, costs were likely still relatively high due to the need for hand painting.[66]

With a wider availability of garden gnomes, resale also began in the 1940s. In a 'Wanted to Sell' column, Wellington's *Evening Post* advertised for sale in 1945 'Dwarf ornaments (2 large, 1 small)',[67] while later that year in a similar column, *The Auckland Star* advertised 'Bird Bath, also colourful garden dwarfs'.[68] Whether this signifies the beginning of a loss of status of the garden gnome is unclear. The garden gnomes at Sutherland's 'Homewood', at least, were still an attraction as late as 1941.[69] However, in a more cryptic note, a lost-and-found notice from late-1943 claims: 'If persons who took Gnome and Frog from 23 Apihai St, Orakei, calls again they may have the mushrooms.'[70] Another theft, but perhaps not an unwelcome one?

Post-1950s gnomes

From the mid-1940s, soldiers started to return from the various theatres of war, and New Zealand enjoyed a post-war boom in ornamental and vegetable gardening.[71] New Zealanders' dream of owning a quarter-acre section — what Austin Mitchell in his 1972 book of the same name called, tongue-in-cheek — 'the half-gallon quarter-acre pavlova paradise' — was encouraged by government through provision of state housing.[72] Societal pressure also valued conformity, expressed through a desire to return to a quiet, pre-war lifestyle. Coupled with conservative societal attitudes, white New Zealand in the 1950s enjoyed some of the highest standards of living in the world thanks to a booming wool industry. Within this conservative culture, gardening assumed great importance as a measure of one's hard work.

Just as increased living standards enabled New Zealanders to afford more things, including more garden implements and figures, so increased technological developments served to decrease the production price of garden gnomes. Most garden gnomes began to be more cheaply produced in factories.

Concrete came to be the dominant material of construction for garden gnomes, and has the advantage of being long-lasting. These societal, technological and economic changes contributed to increasing the popularity of garden gnomes from the 1950s; from this period, garden gnomes were no longer an elite object. Such has been their popularity, that it has been difficult if not impossible to track with any certainty the precise contours of their post-1950s history. Nevertheless, several recent salient points emerge, exhibiting continuities and discontinuities with the period up to the 1940s.

The theme of theft, which began so early in New Zealand, has continued into modern times, which may provide a reason why — despite being readily available — garden gnomes are less commonly visible in front gardens than they once were. Since 2000, regular reports continue to appear in newspapers of gnomes lost,[73] found without owners,[74] or reunited with their owners.[75] For some, as with the disappearance of the garden gnome of Rue Bulow of Stoke, in 2009, they follow trends overseas, where a gnome is whisked away on a holiday adventure, with a note left for the owner of their impending adventure.[76] The note for Mrs Burlow stated: 'I will photo-document these adventures and send to you some fun pictures of this trip. When we have completed this magical journey, your gnome will return home to you.' When three Dunedin students were caught with stolen garden gnomes in 2012, however, they simply claimed to be taking part in a tradition to liberate garden gnomes,[77] and that students take on the role of 'gnome keepers'.[78] The police, however, saw it differently, with Sergeant Wayne Brew of the Oamaru Police stating, 'in reality it's not a tradition, it's theft'.[79]

Especially evident from the 1980s has been the growth in New Zealand of a popular culture around gnomes that has parallels with international trends, yet with important local characteristics. Their persistence and importance in New Zealand culture is indicated by Christchurch's hosting of the 'First International Convention of Garden Gnomes' in 1995. This event attracted around 200 gnomes and eight to ten thousand visitors, and featured a special visit from the earliest surviving English gnome, 'Lampy'. Garden catalogues from the 1980s also suggest there was a wide range of designs available in New Zealand, with one showing 19 different gnomes, including the Disney-style 'seven dwarfs' figures.[80] Alongside the internationalization of gnomes has been the emergence of a distinctive New Zealand gnome culture. At least from the 1980s, if not before, home-grown concrete gnomes were being manufactured domestically, and were even exported from New Zealand to Australia.[81] A popular series of garden gnomes has been that released by the beer producer 'Tui' since 2000,[82] as part of their highly successful branded food and merchandise range.[83] These resin gnomes, with Tui-branded bucket hats, rather than the traditional pointed headwear, continue to be released, and include characters undertaking activities relevant to the age group the beer retailers market to, figures also strongly associated with New Zealand; for example, cricketers, surfers, hitchhikers and, of course, beer drinkers (figure 3). Overall, the inexpensive garden gnomes available today, with their varied construction, are a far cry from the terracotta specimens in 1930s New Zealand.

Garden gnomes have also infiltrated other areas of New Zealand popular culture. They appear in New Zealand children's books,[84] on an episode of the locally produced, globally televised, children's show 'The WotWots', and in Academy Award™ winner Peter Jackson's 1992 comedy-horror film, 'Braindead' (where a decapitated zombie has a garden gnome inserted where his head used to be). Further, they have made their way into high art. For example, in 2013 large gnome statues made of mirror polished stainless steel by sculptor Gregor Kregar were erected outside Christchurch Art Gallery;[85] these were removed in June 2014.[86]

Conclusion

Garden gnomes have interesting stories to tell, in England, New Zealand and likely also elsewhere. Based on the differences observed between the histories of New Zealand and English garden gnomes, and due to their status as cultural icons within the gardens of these nations and others, the histories of these garden figures deserve greater attention. In contrast to England, garden gnomes only appear in the documentary record in the 1930s as elite objects, by which time in Britain the statuary was becoming more popular among non-elite consumers. Economic and technological changes rendered garden gnomes more readily available from the late 1940s and into the 1950s, such that over the last twenty years or more, garden gnomes in New Zealand have become associated with aspects of that country's cultural identity, from their appearance in popular advertisements and books to a comedy horror, and even 'high art'.

But what also of other, less enduring, garden ornaments? What of 'Pink Flamingos', for example, which had a briefer, but still wide ranging appeal? Or what of those that were more regional? In New Zealand, colourful cut-

FIGURE 3.   *A gnome released by the popular 'Tui' beer brand. Source: photograph by author.*

out metal butterflies, attached to the sides of weatherboard houses, decorated the gardens of many suburban homes in the 1950s.[87] Concrete seals, balancing balls on their noses, were also commonly encountered. In contrast, concrete aboriginal figures, known as 'Neville' and 'Noeline', populated Australian gardens,[88] while North America had 'Lawn Jockeys' and concrete geese.

Overall, the history of the garden ornaments of the masses, commonly dismissed as being kitsch or in bad taste, is sadly under-researched globally. The men (and increasingly women) with pointy hats, who stand sentinel over our gardens come rain or shine — and other garden ornaments — deserve much better from garden historians.

Acknowledgements

Thank you to James Beattie for the invitations to present this work at the 'Gardens at the Frontier' conference and to write this work, and for helpful suggestions in preparing this article. I also extend a large thank you to John P. Adam for scanning material from the National Library, and to the reviewers who provided useful comments on this manuscript.

Disclosure statement

No potential conflict of interest was reported by the author.

NOTES

1. Twigs Way, *Garden Gnomes: A History* (Oxford: Shire Library, 2009).
2. Ibid.; Gordon Campbell, *The Hermit in the Garden: From Imperial Rome to Ornamental Gnome* (Oxford: Oxford University Press, 2013).
3. Way, *Garden Gnomes*.
4. Ibid.
5. For example, Helen Leach, *1000 Years of Gardening in New Zealand* (Wellington: AH & AW Reed Ltd, 1984); Matthew Bradbury (ed.), *A History of the Garden in New Zealand* (Auckland: Penguin Books, 1995); Bee Dawson, *A History of Gardening in New Zealand* (Auckland: Random House, 2010).
6. Way, *Garden Gnomes*, p. 40.
7. For example, Mark Stocker, '"Director of the Canoe": The Auckland Statue of Sir George Grey', *Melbourne Art Journal*, xi/12, 2008–2009, pp. 50–63; Mark Stocker, '"Drape the Gross with Grace"? The Auckland Athlete Statue and Its Critics', *Sculpture Journal*, xxi/1, 2012, pp. 118–126.

8. Katherine Raine, 'Early Twentieth Century Gardens', in Matthew Bradbury (ed.), *A History of the Garden in New Zealand* (Auckland: Penguin Books, 1995); Louise Beaumont, 'Gardens of the 1920s and 1930s', in Bradbury (ed.), *A History of the Garden in New Zealand*, pp. 150–151.
9. James Beattie, Jasper M. Heinzen and John P. Adam. 'Japanese Gardens and Plants in New Zealand, 1850–1950: Transculturalism and Transmission', *Studies in the History of Gardens and Designed Landscapes*, xxviii/2, 2008, pp. 219–236.
10. James Beattie, 'Making Home, Making Identity: Asian Garden Making in New Zealand, 1850s–1930s', *Studies in the History of Gardens & Designed Landscapes*, xxxi/2, 2011, pp. 139–159.
11. Angela McCarthy, 'Migration and Ethnic Identities in the Nineteenth Century', in Giselle Byrnes (ed.), *The New Oxford History of New Zealand* (South Melbourne: Oxford University Press, 2009), pp. 173–196.
12. Ibid.

13. Way, *Garden Gnomes*, pp. 19–24.
14. David Tannock, *Rock Gardening in New Zealand* (Auckland: Whitcombe and Tombs, 1924).
15. 'Outdoor Living Room. Novelties for the Garden. Display at John Courts', *New Zealand Herald* (19 March 1931), p. 14.
16. Way, *Garden Gnomes*, pp. 10–15.
17. 'Advertisements', *Auckland Star* (14 November 1931), p. 7.
18. Way, *Garden Gnomes*, pp. 46–47.
19. Janice Mogford, 'Court, John 1846–1933', *The Dictionary of New Zealand Biography* (Auckland: Auckland University Press, 1996), Vol. 3, pp. 118–119.
20. Ibid.
21. According to the Reserve Bank of New Zealand's inflation calculator, a basket of goods and services that cost 29s 6d in late 1931 equates to approximately NZ$150 in 2014 (approximately UK £75 in 2014), while the 99s 6d is over $500 (£250).

22. Philippa Mein Smith, *A Concise History of New Zealand* (Melbourne: Cambridge University Press, 2005), pp. 150–151.

23. 'A Charming Studio', *Evening Post* (1 November 1932), p. 11.

24. Margaret Nell Galt, *Wealth and Income in New Zealand: c. 1870 to c. 1939*, Unpublished PhD thesis, Wellington, University of Victoria, 1985.

25. Jim McAloon, 'Gentlemanly Capitalism and Settler Capitalists: Imperialism, Dependent Development and Colonial Wealth in the South Island of New Zealand', *Australian Economic History Review*, xlii/2, 2002, pp 204–233.

26. 'Garden Parties for Free Ambulance: Weather Intervenes', *Evening Post* (7 March 1932), p. 3.

27. Beryl Smedley, *Homewood and Its Families* (Wellington: Mallinson Rendel, 1980), p. 103; Rupert Tipples, *Colonial Landscape Gardener: Alfred Buxton of Christchurch, New Zealand 1872–1950* (Lincoln College, Christchurch, 1989), pp. 101–106.

28. Smedley, *Homewood*, pp. 101–110.

29. Beattie, Heinzen and Adam, 'Japanese Gardens and Plants in New Zealand'.

30. 'Garden Parties for Free Ambulance: Weather Intervenes', *Evening Post* (7 March 1932), p. 3.

31. 'Homewood' Grounds Open Sunday Afternoon', *Evening Post* (13 March 1941), p. 2.

32. 'Homewood's Ferneries', *Evening Post* (13 March 1931), p. 14.

33. Pamphlet, in Smedley, *Homewood*, pp. 106–108.

34. Diana Beaglehole, 'Sutherland, Benjamin 1873–1949', *The Dictionary of New Zealand Biography* (Auckland: Auckland University Press, 1998), Vol. 4, pp 505–506.

35. John Stacpoole, 'Mappin, Frank Crossley 1884–1975', *The Dictionary of New Zealand Biography* (Auckland: Auckland University Press, 1998), Vol. 4, pp. 333–334.

36. 'An Auckland Garden: The Joy of Flowers', *New Zealand Herald* (26 September 1932), p. 6.

37. James Beattie, 2014, personal communication.

38. 'Advertisements', *Evening Post* (27 February 1935), p. 1.

39. Tim Shoebridge, *An Unrivalled Private Residence: A History of 186 The Terrace and Its Occupants, 1839–2000* (Wellington: T. Shoebridge, 2000).

40. 'Advertisements: Stolen: Reward £5', *Hutt News* (13 December 1939), p. 8.

41. Way, *Garden Gnomes*, pp. 12–13.

42. For example, 'Hutt Valley Lierdertafel', *Hutt News* (16 August 1933), p. 4.

43. 'Advertisement', *Evening Post* (11 February 1937), p. 27.

44. Way, *Garden Gnomes*, pp. 50–53.

45. 'Rock Garden Society: Annual Meeting', *Hutt News* (30 June 1937), p. 4.

46. Diana Beaglehole, 'Gibbons, Hopeful 1856–1947', *The Dictionary of New Zealand Biography* (Auckland: Auckland University Press, 1996), Vol. 3, pp. 171–172.

47. 'Auctions', *Evening Post* (24 January 1940), p. 16.

48. 'Disappointing Weather', *Evening Post* (28 February 1938), p. 16; 'Garden Party', *Hutt News* (22 February 1939), p. 5.

49. 'Public Auction', *Evening Post* (17 November 1939), p. 14.

50. Rupert Tipples, 'Buxton, Alfred William 1872–1950', *The Dictionary of New Zealand Biography* (Auckland: Auckland University Press, 1996), Vol. 3, pp. 82–83.

51. Ibid.

52. Beattie, Heinzen and Adam, 'Japanese Gardens and Plants in New Zealand'.

53. Beaumont, 'Gardens of the 1920s and 1930s', pp. 150–151.

54. Ibid.

55. Way, *Garden Gnomes*.

56. Stacpoole, 'Mappin, Frank Crossley'.

57. Graham W. A. Bush (ed.), *The History of Epsom* (Auckland: Epsom & Eden District Historical Society Inc., 2006), p. 406.

58. Shoebridge, *An Unrivalled Private Residence*.

59. James Belich, *Making Peoples: A History of the New Zealanders from Polynesian Settlement until the End of the Nineteenth Century* (Auckland: Penguin, 1996), pp. 328–332.

60. Guy H. Scholefield, *Newspapers in New Zealand* (Wellington: A. H. & A. W. Reed, 1958), p. 23.

61. Including 'Wanted to Sell', *Evening Post* (8 May 1940), p. 1, among others.

62. 'Garden Needs', *New Zealand Herald* (5 June 1940), p. 4.

63. 'Smith and Smith' Advertisement, *New Zealand Herald* (10 December 1943), p. 3.

64. *Griffiths Gardening Spring 1946* catalogue (Auckland: Griffiths Ltd., 1946).

65. *Griffiths Gardening Winter 1947* catalogue (Auckland: Griffiths Ltd., 1947).

66. Way, *Garden Gnomes*, p. 35.

67. 'Wanted to Sell', *Evening Post* (11 April 1945), p. 3.

68. 'Wanted to Sell', *Auckland Star* (21 December 1945), p. 2.

69. 'Homewoods Ferneries', *Evening Post* (13 March 1931), p. 14.

70. 'Lost and Found', *Auckland Star* (2 November 1943), p. 1.

71. Helen M. Leach, 'Exotic Natives and Contrived Wild Gardens: The Twentieth-century Home Garden', in Eric Pawson and Tom Brooking (eds), *Environmental Histories of New Zealand* (Melbourne: Oxford University Press, 2002), pp. 214–232; p. 223.

72. Austin Mitchell, *The Half-Gallon Quarter-Acre Pavlova Paradise* (Christchurch: Whitcombe and Tombs, 1972).

73. 'Disgraceful Theft', *Auckland Now* (27 January 2009), no page.

74. 'Gnomeless in Carterton', *Dominion Post* (30 July 2008); 'Police Take a Dim View of Gnome Prank', *Manawatu Standard* (3 November 2009).

75. 'Woman United with Little People', *Manawatu Standard* (6 November 2009), no page.

76. 'Gnome Matter; He'll Surely Come Back', *Nelson Mail* (6 February 2009), no page.

77. 'Plea for Gnomes to Come Home', *Otago Daily Times* (14 April 2012), no page.

78. 'Go Gnome, Stay Gnome', *Critic* (April 2012), no page.

79. Ibid.

80. Catalogue for 'Garden Ornaments Ltd', Auckland, 1985.

81. 'Doing Something Concrete', *New Zealand Listener* (5 November 1983), pp. 36–37.

82. 'Tui Gnome', *Marketing Magazine* (NZ) (November 2003), no page.

83. 'Tui Barrels Ahead with Million-Dollar Sideline', *Dominion Post* (5 January 2009), no page.

84. Donovan Bixley, *The Looky Book* (Auckland: Hachette New Zealand Ltd, 2012).

85. 'City Gallery Lets Its Art out to Play', *The Press* (18 May 2013), no page.

86. 'Sculptures Moved to New Gnome', *The Press* (27 June 2014), no page.

87. Graham Hutchins, *Strictly Kiwi* (Auckland: Hodder Moa, 2010).

88. 'Bring Back Neville', *Newcastle Herald* (28 September 2009), no page.

Index